Communicating Science

Pierre Laszlo

Communicating Science

A Practical Guide

Prof. Pierre Laszlo
„Cloud's Rest"
Prades
F-12320 Senergues
France

DOI 10.1007/75951
ISBN-10 3-540-31919-0 Springer Berlin Heidelberg New York
ISBN-13 978-3-540-31919-1 Springer Berlin Heidelberg New York
e-ISBN 3-54031920-4
Library of Congress Control Number: 2006920778

Springer is a part of Springer Science + Business Media
springer.com

Coverdesign: design & production GmbH, Heidelberg
Typesetting & Production: LE-TEX Jelonek, Schmidt & Vöckler GbR, Leipzig
Printed on acid-freepaper 2/YL – 5 4 3 2 1 0

To the memory of Lewis Thomas (1913–1993),
who set an example to all of us
as scientist, communicator and humanist.

PREFACE

Many years in the making, this book aims at improving the skills as communicators of my fellow-scientists. Advancement of knowledge is synonymous with diffusion of results, first within the scientific community and then to the public at large. Accordingly, the book has two main parts, corresponding to these two constituencies, with their widely different needs. A third, smaller section is devoted to informing decision-makers. I have attempted a rather comprehensive coverage.

Readers whose native language is not English – which incidentally is also my case – may find the book useful. While this is not a manual on speaking and writing English – there are plenty of those around – they may find it educative nevertheless.

In the vast majority of cases, I have first-hand experience of what I write about. In addition, I have consulted a recognized expert on each topic.

Each of the main parts is subdivided into two sections, Guidelines and Genres. The segments are set in alphabetical order within each of those subsections. Hence, the table of contents serves at the same time as an index. Within each segment, a few cross-references remind the reader of related subject matter in other parts of the book. Inevitably, there are repetitions, for which I apologize. A brief bibliography of some useful articles and books is provided.

It is my pleasure to thank for their precious advice, which has markedly improved this manual, my friends and colleagues: Philip Ball, Alex Bellos, Jim Bennett, Wesley T. Borden, John Hedley Brooke, Gregory L. Diskant, Roald Hoffmann, Jamie C. Kassler, Joseph B. Lambert, Robert L. Lichter, Annette Lykknes, Jozef Michl, Peter J. T. Morris, Guy Ourisson, Philippe Perez, Daniel Raichvarg, Christopher Ritter, William Roberts, Edward T. Samulski, Jeffrey Seeman, Peter Stang, Mel Usselman, Laura van Dam, Stephen J. Weininger.

I was gratified by the impressive professionalism of Dr. Marion Hertel and her colleagues at Springer, in Heidelberg.

My thanks go, last but not least, to my wife Valerie. She has gratified me with her skillful and imaginative editing of the individual pieces, making them both leaner and tastier.

Spring 2006 Pierre Laszlo

CONTENTS

PART I

ADDRESSING PEERS

GUIDELINES

GENRES

PART II

THE GENERAL PUBLIC

GUIDELINES

GENRES

PART III

DECISION MAKERS

GENRES

PART I

ADDRESSING PEERS

GUIDELINES

ABSTRACT

An abstract is like the face of a person. It can tell one what to expect. It should give the gist of the paper in a short paragraph. Besides being a summary, it has another purpose. A showpiece, it beckons the reader into the paper. You do not want to compose an abstract so well devised the prospective reader after glancing at it will decide to skip the paper. But even worse, you want to avoid writing an abstract so discouraging as to turn the reader away from both the abstract and the paper itself.

Don't compose an unreadable abstract. This would seem to go without saying. An opening sentence such as »*Fragments of polyketide synthase (PKS) genes were amplified from complementary DNA (cDNA) of the fusarin C producing filamentous fungi Fusarium monoliforme and Fusarium venetatum by using degenerate oligonucleotides designed to select for fungal PKS C-methyltransferase (CmeT) domains*« is impossible: Too technical, too dense, too complicated a syntax, heavy with **acronyms**.

Let's start with translating and breaking down this sentence. How about, instead, »*Two species of filamentous fungi produce fusarin. Their complementary DNA serve to amplify fragments of polyketide synthase genes. We use for this purpose degenerate oligonucleotides, designed to select for the desired methyltransferase domains.*« Moreover, since the **title** of the paper starts with »Fusarin C Biosynthesis«, don't repeat it. The desired meaning of the above sentence is »*We have prepared oligonucleotide PCR primers selective for fungal polyketide synthase genes.*« Why not start the abstract thus?

Think of an abstract as a shop window. It requires elegance and attractiveness. The latter ought to be a reflection of the quality of the work, of the novelty of its approach, of the importance of its results. The former is exclusively a matter of word-craft, of style. It is not enough to whip out an abstract in five minutes, either before writing the paper or after having done so. A good abstract might be hours in the making. To invest time assembling such a jewel is not out of whack.

How can it be done? Write as if you were penning a postcard to a friend: use simple sentences, don't get technical and utter a clean message in a maximum of 50 to 100 words.

Incapable of such a feat? In that case, take a printout of your completed paper and underline a dozen sentences you feel epitomize the work. Paste them together and tie them together with transitions. Now, edit this paragraph. Be merciless. Try to reduce it by a third. You should now have the first draft of your abstract. It only remains to turn it from decent into artful English. Do not hesitate to resort to a dictionary of synonyms and to use other tools such as a style manual. Read your abstract aloud, a crucial test. It will make you jettison multisyllabic unpronouncable words. It will make you focus on the genuine achievements of the paper. It will make you grab your reader by the sleeve: »*come inside, Mister, I have something truly marvellous I'd like to show you*«. This is the message from a well-designed abstract.

REFERENCE: K. K. Landes (1951) *Scrutiny of the abstract.1.* AAPG Bulletin, 35(7):1660; (1966) *Scrutiny of the abstract.2.* AAPG Bulletin 50:1992.

Whom to thank? 

ACKNOWLEDGEMENTS

This appears an almost perfunctory and ancillary part of a publication. Such a perception is misleading.

This segment is worthy of your full attention. Failing to thank someone for assistance does not speak highly of your accuracy and care – two essential qualities in a scientist.

Which brings up the question: whom to thank? The answer could not be simpler: anyone who is not listed among the authors and who nevertheless has somehow helped or contributed. The criterion for authors, covered elsewhere in this book, is very simply that any of the co-authors ought to be able to present and defend the paper, as a whole or in part.

How then should this segment be presented and written? There is no prescribed format. Nevertheless, I shall suggest one. Why not take model on the Acknowledgement section of nonfiction books? It is often prepared and written with great care. The author uses this opportunity to recapitulate the history of the book and to set it on the record. By doing likewise with your text, you will be able not only to thank all the relevant parties, but also to put together a piece of writing which, in the future, historians may turn to as a source of valuable information.

REFERENCE: R. A. Day (1994) *How to write and publish a scientific paper.* 4th edn, Oryx Press, Phoenix AZ.

They are eyesores. 

ACRONYMS

The TA is BI. This assertion, which could be construed as stating the sexual preferences of a teaching assistant (TA), is meant here as »the topic of acronyms is a bothersome issue«. One of the many problems with acronyms is the multiple meanings many of them have. Of course, within the confines of a single publication, addressed primarily to a group of specialists, such ambiguities vanish.

Acronyms are necessary due to the need for brevity. Long names of chemicals beg to be shortened. Thus monopyrrolotetrathiafulvalene (29 letters, far from a record) is shortened into MPTTF, with only five. Some such chemical acronyms have entered common language. Examples include DNA, RNA, TNT, DDT, … Names of commonly-used laboratory tools and techniques are also turned into acronyms. Fourier-transform infrared spectrometry becomes FTIR (four instead of 37 letters), atomic-force microscopy is AFM and a superconducting quantum-interference device is a SQUID.

This last example illustrates a consequence of acronyms entering the scientific lingua franca. A naïve student using the SQUID and who continually hears it referred to as such may never learn what the acronym stands for. To him, the acronym is an opaque screen. He does not know the real composite name hiding behind, hence he may never find out how the contraption works. Is it such a hot idea for a user not to understand how an instrument works?

But acronyms have some redeeming value, too. Their very existence points to the pressure for efficient communication within the scientific community. Indeed, acronyms help to network research scientists worldwide. Often, an in-house abbreviation within a single research group transfers into common scientific language.

Another nice aspect of acronyms is their testifying to the playfulness of scientists. Play is an essential component of science. To make-up an acronym, a wordplay, belongs to such an attitude.

Do's and dont's? My first rule is to try and stick to a maximum of three letters in a non-punning acronym, such as DNA or TNT. One of the reasons is that, in a language such as English, it is difficult to pronounce more than three consecutive consonants. If you need more letters, then you should make sure that you have enough of a balance between vowels and consonants for easy pronunciation. A common trick is to tailor the acronym to an already existing word, as exemplified by SQUID. Quite a few scientists have had a field day turning loaded words into acronyms, such as PENIS standing for an algorithm used in nuclear magnetic resonance (NMR for short). Make sure that, whenever an acronym first appears in a publication, it is flanked with its translation. But make as little use of acronyms as possible. They clutter a text, they hinder the flow of reading, they are eyesores and tongue-twisters, they are a necessary evil.

REFERENCE: K.T. Hanson (1995) *The art of writing for publication*, Allyn & Bacon, Needham, MA.

switch from the passive to the active 

ACTIVE OR PASSIVE VOICE?

This question is a dilemma only because the conventional way of writing scientific papers heavily favors the latter. But first, let's give a look at a few samples of such writing.

»*The characteristics of … have been suggested so far to be the result of cooperative phenomena.*« The timidity in this sentence reflects a very real and respectable uncertainty, in attribution of a cause to scientific results. It stems from the difficulty of interpretation, which can never be certain and which, therefore, has to be muted. But there is a fine line between such prudence and downright obfuscation. One wishes the author of this paper had stuck his neck out and written for instance »*Cooperative phenomena account for the characteristics of …*«. If this were too strong an assertion, it could be qualified: »*Cooperative phenomena account, at least in our view, for the characteristics of …*«.

Another example: »*This work was motivated by previous demonstrations that …*«. This opening statement aims at identifying a historical continuity, the study we are about to read did not arise in a void, it followed upon earlier work. Why not write instead »*We base this work on previous demonstrations that …*«, which requires only a rephrasing from the passive to the active voice? Notice that, not only did I turn the sentence from passive to active, but I also moved it from the past tense to the present.

The habit of using the past tense is another feature in writings by scientists. When colleagues pen this sentence: »*Support effects have been attributed to co-reactant activation sites …*«, thus referring to an opinion which they might wish to challenge, why don't they write instead »*Do co-reactant activation sites truly produce support effects?*«

Sometimes, the combination of a switch from the passive to the active, together with dispensing with unnecessary words, works quite nicely: »It was demonstrated that this surface provides the best efficiency for ...« thus easily becomes »This surface is most efficient at ...«.

Turning to the passive voice is second nature to scientists. This is their way of playing possum. It is a diversion tactic against possible criticism of their work by referees and peers. They qualify their statements making them innocuous in the hope of deflecting any challenge. In so doing, assertions lose their edge, the writing becomes moot and the reader's attention wanders.

REFERENCE: M. Young (1989) *The technical writer's handbook*, University Science Books, Mill Valley, CA.

your paper has three constituencies 

BIBLIOGRAPHY

Why is a good bibliography essential? It demonstrates your seriousness as a scientist. What is a good bibliography? It displays honestly and conscientiously the foundations of your work. The phrase »scholarly apparatus« is synonymous with a bibliography: the end part of your paper is a testimony to your **scholarship**. Your results may be trusted, you are also a scholar to be respected. You know what you are writing about, you know your place in the science world, you read what others have published in the field.

The goal of a list of references is not one-upmanship. Avoid being vain, because it will invite ridicule and scorn, as if you were wearing a flashy suit with a gaudy tie and outrageous multicolored shoes. Refer only to those of your earlier publications truly critical to the understanding of the present work. Nor is the sole purpose of a bibliography to ingratiate yourself with the most influential among your peers, those most likely to be called upon to serve as reviewers or referees.

Your paper has three constituencies among its readership. Your peers make up only one. No less important are the young scientists, those starting in the profession, the graduate students. Make your paper helpful to them. Select for their didactic value three papers, or book titles, for inclusion at the very beginning of your list of references. The third constituency you ought to keep in mind is that of present and future science historians. They need to configure the network of scientists active in a field. Inclusion of not only your friends but your enemies, is a must. You want your paper to have lasting value? Then, rather than omitting any but the most recent references, those published during the last couple of years, provide a decent time line.

At the risk of stating the obvious, the first requirement for a bibliography is total reliability. Check and recheck against the original publication in the journal, the spelling of the names, the volume and page numbers. You will be surprised by the number of errors which can and do occur in transcription.

If you have simply lifted your bibliography from another paper, such errors will also broadcast to the world the cardinal sin of a copy-and-paste job – this is plagiarism. It is actionable. If you do not have access to a library and thus are unable to read publications in your field, ask for the cooperation of a colleague elsewhere, operating under wealthier circumstances. He or she will be happy to give such help to a fellow-scientist.

A way to check whether you have provided a fine bibliography is to be prepared to give, at the drop of a hat, a summary of each cited paper.

REFERENCE: J. S. Dodd (ed) (1997) *The ACS style guide*. American Chemical Society, Washington DC.

It suggests availability. 

BODY LANGUAGE

We scientists may be polyglots, able to read and write, and even speak several languages. We have some mastery of the language of mathematical equations. We are fluent in the language of the scientific discipline we have specialized in. But there is one particular language at which, as a rule, we are poor. That is body language.

The stereotypes are aired on movies, on television and in cartoons. While they exaggerate, they carry some truth. The bespectacled scientist is seen as an unattractive middle-age male who looks rigid, boring and uncompromising. He is visibly not at ease. His body lacks grace and charm. He cannot dance.

Counter those stereotypes. You love your science. Show it, express your enthusiasm bodily. Be dynamic whenever you are engaged in any form of science communication. Instead of standing stiffly behind a lectern or while pointing at an image on a screen, move about the room. Get close to the listeners, make eye contact with them individually. Meet people with an engaging smile. To smile is to grant a welcome. It suggests availability. It tells people without words that you are open and more than willing to share your knowledge and worldview.

The way you dress sends, not a single signal, a whole alphabet. This is too obvious to elaborate upon. Look at yourself in the mirror, ask a family member to help you in selecting your attire (attire as in »attraction, to be attractive«). The way you hold yourself, your involuntary gestures – a leg moving rythmically, a tic with your hand, your furrowing your eyebrows, etc. – all assume significance and can be detrimental to your message.

You can improve in those sectors by regular physical exercise such as walking, swimming, tennis or gymnastics. An instructor or personal trainer in physical exercise surely would help. Likewise, consulting a physical therapist. Better yet, you could hire a coach who specializes in preparing public speakers. This person can also help you greatly by monitoring your voice, controlling its pitch and volume, and directing it at specific parts of your audience, making it carry drive and feeling. What your voice will express, your body will also start to follow. The voice can lead the rest of the body in expressiveness.

Starting with your hands. They are your most precious auxiliary. You can learn to make better use of them, in the following way. Stand in front of a mirror and read aloud, for instance a **paragraph** from a newspaper. Now repeat the performance and make sure to use your hands, in order to highlight points and create emphasis. Do this for a couple of minutes twice a day for a week. You should notice an improvement, your hands now accompany your speech. The second week, replace the newspaper with one of your own texts.

A final point: talking with your hands, more generally using body language will help you to act out the invisible. Your work involves characters and forces, perhaps at the microscopic level, perhaps in the astronomical range, in any case outside normal perception. It will help your listeners if you convey your mental image of such actors in your work with a little gesturing. A little: remain somewhat restrained and discreet, and you will be all the more convincing.

REFERENCE: D. B. Givens (2005) *The nonverbal dictionary of gestures, signs & body language cues*. Center for Nonverbal Studies Press, Spokane WA.

Wrapping-up your story 

CONCLUSION

When a letter is written, it goes into an envelope. When the mail arrives, we can usually tell from the mere aspect of the envelope, whether it contains a bill to be paid or a letter from a dear friend. The conclusion to an article published in a scientific periodical has this same feature, it qualifies the paper as to its type, its merits and its lasting value. Keep it in mind when drafting a conclusion, do not treat is as perfunctory. Do not make it a study in conformity, nor a repeat of the **abstract** of your paper. Wrapping-up your story is a significant component of your publication.

A good conclusion should be eloquent and it should remain modest. Eloquent, because you want to leave your readers on an upbeat, so that they hold a good impression. And yet modest, do not regale yourself with what you have just accomplished: how pioneering, how outstanding an achievement! Blowing your own horn is a turn-off.

The optimal length of a conclusion is a couple of paragraphs, no more than a page overall, i.e., about 1,500 characters at most. And what are, or should be its ingredients? The Discussion part of the paper will have already critically examined the foundations of the work, the key assumptions made, the quality of the experiments, the statistical significance of the measurements, …, so there is no need to reiterate this in the Conclusion. Towards the end of it, readers should find a one-sentence summary of the work, couched in a neutral, objective manner – remember to avoid one-upmanship, it stinks. Readers need such a sentence to reassure and perhaps convince themselves they have grasped the essence of what you had been telling them.

Bringing a scientific publication to its conclusion need not resemble the closing of a door, for two reasons: any scientific contribution is open-ended, moreover it is open to criticism and controversy too. You wish to craft your conclusive sentences accordingly. Why not devote them, instead of burning incense to your own immortal glory, to a forward, prospective look?

Thus I recommend that your conclusion consist predominantly of statements about possible or probable extensions of the study which has just been reported upon. It will help you, in any case, to clarify in your mind what this future task consists of, whether you intend to do it or not. It will give your readers the sentiment that you are behaving responsibly, that you have an authorial attitude towards your work, worthy of respect. The clarity with which you will set your results in context, within the evolution of the discipline, more than within the narrower framework of your own evolving research, is a marker of a fine, thoughtful conclusion.

REFERENCE: M. Alley (1996) *The craft of scientific writing.* 3rd edn, Springer, New York.

favor nonprofit organizations 

ELECTRONIC PUBLISHING

Are we witnessing the demise of printed journals, with the transition to electronic publishing? Broadcasting science on the World Wide Web has many advantages. It is less costly, at least in principle. It spares considerable numbers of trees from being turned into paper pulp. Each page appears exactly as it was formatted, from use of software such as the Acrobat pdf. Bypassing printers accelerates publication. Moreover, closeness to e-mailing amounts to a return to the origins: in historical terms, science journals originated with private correspondences between natural philosophers, during the seventeenth century.

Those are all assets. But there are issues too, which need addressing. Free access has been a universal rule for the **Internet**. While consistent with the scientific ethic, which encourages the widest discussion and criticism, it runs somewhat counter to the need of the publishers, including professional societies. These want to make a profit, enough of it in any case to support electronic publishing. One can only hope that the Internet phenomenon will favor nonprofit organizations with respect to commercial publishers. This may well turn out to be a vain hope. Indeed, one should recognize a pivotal consideration. The editorial process bears the major portion of the cost of publication – an invariant in going from traditional to electronic publishing.

One might argue indeed that the main issue of electronic publishing is editorial control. Academic publishing has thrived, over several centuries, by certifying a paper's validity in a number of ways, not only peer review but also prestige of the journal, and the whole scholarly process, from recording an observation in a laboratory notebook to having one's paper see the light of publication.

In the Age of the Internet, peer-review is made somewhat easier and faster by technology. It is essential that the quality control it ensures retains its excellence and be not debased. Scholarly values are inherently conservative, which may hinder at times the spread of revolutionary new ideas. Adherence to them may also stifle participation from developing countries to the advancement of knowledge. Editorial control is necessary at all stages of scientific communication, in order to ensure maximum discussion of the contents of each paper, while protecting its authors from *ad hominem* attacks and from gratuitous controversy, merely for the sake of being contentious.

The Web brings with it a facility, the electronic library. In the near future, each scientific paper will carry a bibliography of so many links, each of which will offer a virtual entry into a subset of the literature. I deem most promising this new dimension of publishing. With the appropriate technology, it ought to vastly decrease the amount of unnecessary duplication of effort and results.

The Web also carries with it a patchwork organization. It is segmented into discussion groups, centers of interest, …, i.e., into a collection of highly specialized niches. This is worrisome. Science needs interdisciplinarity, it also needs for its sociology to include cores of people who are generalists, not specialists. Thus, I encourage you to publish in hyperspace in the less rather than in the more specialized journals. Otherwise, self-selecting of the readership into groups of narrow specialists may give the kiss of death to science. Even a cursory look at science history shows that real progress has always been both unpredictable and on the margins, rather than occurring in the mainstream.

I shall merely mention the issue of copyright, since fighting unauthorized copying and piracy conceivably will make effective use of technological tools. A more worrisome issue, I believe, is the archival: no worldwide agency is presently empowered and funded to preserve all the material which appears on the Net, electronic publications in particular. Yet, scholarly work demands being preserved, and even the more so in the absence of current interest.

Moreover, rapid technological change ensures that supports for the data have an effective lifetime measured in just a few years. The example of floppy disks, Zip disks, CDs and DVDs, comes to mind. Fortunately, for the last 30 years or so, migration of digital documents from a machine, an operating system or an application program has been the preservation method of choice. It may well be so ingrained by now into our mental habits as to become a universal rule for archival safeguarding.

REFERENCE: G. P. Schneider, J. Evans and K. T. Pinard (2005) *The Internet illustrated introductory.* 4th edn, Course Technology, Boston M A.

harbinger of scientific collaborations 

E-MAILING

E-mail is easy, too easy. To e-mail indiscriminately, in mass, is to abuse the recipients, thus also drowning information in noise. The ability to dash off a note to a correspondent from anywhere, by typing on a keyboard, is a dream come true.

The greatest asset of e-mail is arguably its informality. In the office, e-mail has become the choice mode of communication among coworkers, using an intranet web. It has supplemented, to some extent has replaced conversations at the bench, in the instruments room, in the halls or next to the coffee machine.

To turn from local use to the global, e-mail allows one to reach out across the oceans, with no hindrance from the differing time zone – a superiority over phone or fax. This asset has been turned into habit, one wakes up in the morning to a detailed answer. It is such a nice and easy way to exchange information that it is a harbinger of scientific collaborations. I wrote a whole book in collaboration this way, trading chapters (or rather scenes in a play) by e-mail. The two of us did not need to get together at all. Nevertheless, we went through a baker's dozen of revisions.

Beware of the disadvantages of the informality. Those include the shapelessness of quite a few messages, the mispellings, the abuse of abbreviations and of cyberspeak. They blemish the appearance. The contents of a message suffer also from flippancy, if not from downright vulgarity.

E-mail can stoop to the level of a graffito scribbled on lavatory tiles. This need not be, but is too often the case.

E-mail suffers also, not so much in principle but in practice, from the lack of a permanent record. We exchange a large amount of correspondence and it vanishes into thin air. Usually, after a few weeks, months or years, instead of being archived, it is deleted from a personal computer. The loss is definitive. It is grievous, at least to librarians and to historians.

Which serves to emphasize that any message has to be significant, in order to rise from the noise, let alone endure. At the time of writing (autumn 2004), the signal-to-noise ratio for e-mail is of the order of one in ten. For one significant message, we receive on the average about ten pieces of garbage – known as spam in cyberspace. Avoid like the plague spamming your colleagues, or giving them impression of being deluged (by you!) with information they can do without. In that category: militant political propaganda, commercial advertising, newspaper clippings passed on, cartoons and jokes, … This is my recommendation, resist proselytism, i.e., curb the impulse to widely circulate unsolicited files to people on an e-mail list, who belong to the circle of your professional acquaintances.

A positive recommendation now, as forceful as the warning just uttered. A good use of e-mail will hone your personal expression. Make it a genuine voice-mail: it is ideal for transmission of your own voice, with its idiosyncrasies, its mannerisms, its most endearing features. E-mail is or can be a written transcription of speech. In-between text and speech, it can become a close approximation to speech, this is up to you.

Truly, e-mail is speech: all the more reason to keep it articulate, direct and clean.

REFERENCE: J. v. Emden (2001) *Effective communication for science and technology.* Palgrave Macmillan, London.

a keen awareness 

ERUDITION

Some people may be surprised to find this word and notion discussed here. To them, erudition evokes old dusty volumes molding on library shelves, an arcane or disused knowledge absolutely antinomic to living, throbbing science. The two main reasons for their myopia are the cults of novelty and of facts. People interested only in facts pay no attention to their progenitors, nor to the process through which such facts were garnered. And yet we know (indeed for a fact) that any scientific fact is theory-laden. In other words, it stands and it means something only within the framework of a theory.

But what is erudition? The function of science is to acquire new knowledge. Another process is the accretion, accumulation and archiving of the existing knowledge. This storage function is performed by libraries, data banks and, most important, in human brains. Erudition is the retrieval of crucial information from such vaults. It is responsible for attribution of intellectual work to its sources. Erudition is the routine, normal activity of the scholarly mind when it delineates the historical roots of things and concepts.

Thus it becomes very obvious why erudition pertains to science and its communication. Science is cumulative. It is a layered manifold. Any understanding, whatever the phenomenon studied, requires leafing through these layers and pages.

Any piece of work builds on earlier knowledge. The authors of a scientific paper need a keen awareness of the earlier work which their own study has built on. Not only do they need to assemble such a **bibliography** in the course of their work, they also need to acknowledge in print their indebtedness to the earlier authors. It is part of the evidence which they have to provide for their readers. In other words, erudition is the means by which they will make their own contribution stand out. Only the historical dimension supplies such a background, puts a piece of work in perspective and shows it as either epochal or trivial.

The **Introduction** to your paper should carry this information. Supply the needed historical continuity between your work and that of your predecessors. The following image is not too far-fetched: only erudition will enable you to build a stand upon which to set your results. Not only is this the only way in which they will make sense, it may well provide them with a pedestal for public admiration.

Reference to the past and even to the distant past anchors your publication in history. Instead of belonging with the flotsam of debris which the next wave, the next scientific fashion will sweep away and disperse, you will inject your contribution with lasting value.

With a little luck, erudition may enrich your work with much more, with something essential. You may uncover a missing link which everyone until now had overlooked. It will put paid to the notion (feeble) that anything more than a few years-old (three years? five years?) can safely be discounted and is best ignored.

REFERENCE: M. Turner (1998) *The literary mind*. Oxford University Press, New York.

an understanding at a single glance 

FIGURES AND CAPTIONS

Make sure that each figure be essential to your paper: essential to the paper, not to your pride. As soon as you realize that you are inserting a figure to advertise the talents in your research group, instead of illustrating a point in the argument, remove it. It is extraneous. It is dead wood. Ultimately, it detracts from your article because it impedes the flow.

What makes a figure indispensable? First, it is a shorthand, a shortcut as well. As the Chinese proverb puts it, a figure may be worth a thousand words. Second, it needs to be attractive. The graphics need to be clean and demonstrative. Find a good depiction of the data points so that their plot makes a nice, interesting curve. Thirdly, a figure may convey information in a unique manner. It may be a need to depict schematically a piece of apparatus, or a model for the data, a mechanism, part of your argument, what have you. In such cases, a figure is indeed the only option.

Whenever your figure is a graph of the data, do not plot out the whole set of data points. Restrict yourself to the main pieces of data. Easy readability of the figure is the reason for such a rule. It is easy to remember in terms of Edward Tufte's Least Ink Principle: any table or figure in a paper ought to be at its most economical of the printer's ink, if it is to be grasped easily, by the eyes and by the brain. Such a cleanup amongst the plotted data is a form of respect for your readership.

Insert a particular figure both within the appropriate **paragraph** – this is not necessarily self-evident, your first decision may not be the best, hence this choice of location deserves careful consideration and even debate within your research group – and as a milestone along the road travelled, i.e., the whole sequence of figures; some readers may well look over the figures before they start reading the paper. The graphics in your paper are an integral part of it.

Yet, at the same time they need to be semi-autonomous from the text.

They tell their own story. Think of their collection as equivalent to the **abstract** of the paper, or to the summary in case you end the paper with a summing-up. Nowadays, many journals advertise their wares with a graphical table of contents. Think of the set of figures in like manner, as a kind of advertisement, informative with respect to what's in the paper and giving the gist of it.

Yes, our colleagues are submerged with information of every kind. Make it easy for them – however humbling to your self-regard as an outstanding prose writer. Offer them an understanding at a single glance (well, at a few really) of the main points of your work.

What constitutes then a successful caption to a figure or scheme? Concision: two sentences maximum. Utter clarity and precision. 100% redundancy with the figure, bringing in any extraneous element is a no-no. 0% redundancy with the text of the paragraph the figure is embedded in.

REFERENCE: Council of Biology Editors (1994) *Scientific style and format: the CBE manual for authors, editors, and publishers.* 6th edn, Cambridge University Press, New York.

claiming priority 

FREE ACCESS

Should electronic publishing be kept unrestrained to individual scientists? This current question may soon be answered. In the meanwhile, it is worthwhile to ponder it, so as to anticipate some of the changes about to occur in science publishing.

Given that the **Internet** is synonymous with a free, unimpeded flow of information, free access is a logical demand. One may draw upon the precedent of blogs: anyone may post on the Net a combination of news and editorials.

Is it however realistic to expect free access to be granted so that any scientist will be free to post his/her results for the purposes of discussion, and of claiming priority and archiving?

Is there to be continued peer review of the material published electronically? Then, the process has to accommodate it as the compulsory first step. In the present printed journals system, peer review and preparing a manuscript for publication are not free. Such steps, with the editing involved, amount to about 50% of the publishing costs in a scientific journal, whether issued from a professional society or a commercial publisher. Transfer from hard copy to electronic publishing won't affect such fixed costs. Who is to pay? Authors are notorious for declining to foot the expense of publication. Even with a change in mentality, this factor alone puts a dent in the ideal of open access.

Moreover, free access as a concept clashes with the present realities of scientific publishing: is scientific research more an egalitarian or an elitist enterprise? Open access assumes that it can be made egalitarian. This would be a revolutionary change. Scientific publication currently obeys an elitist, not an egalitarian model.

Bibliometric data are eloquent. They follow Zipf or Pareto-like distributions, which points to a hierarchical social organization of scientific research. In any given field, the square root of the total number of active scientists is responsible for half of the total output. Assume for instance a field with 100 active participants. Only 10 of those sign half of all publications.

Why such a hierarchical structure and where does it originate? Science is inherently conservative because of its cumulative nature. Newly established knowledge piles on top of prior results.

The training of young scientists, who serve an apprenticeship with a master or mentor, is akin to belonging to a guild of craftsmen. Moreover, universities, libraries, doctoral dissertations all share transmission of the existing knowledge among their roles. This, often a primary requirement, make all such institutions inherently conservative.

To sum up: quality control and open access are mutually exclusive. This jaundiced conclusion may be made void by the practices which will develop, I may have overlooked a crucial feature in my analysis. In which case, it will be very interesting to watch the whole new course which scientific publishing finds for itself.

REFERENCE: S. Bachrach, R. S. Berry, M. Blume, T. v. Foerster, A. Fowler, P. Ginsparg, S. Heller, N. Kestner, A. Odlyzko, A. Okerson, R. Wigington, and A. Moffat (1998) *Intellectual property: who should own scientific papers?*. Science 281(5382):1459–1460.

a fundamental opposition 

IDEOGRAPHIC VS. NOMOTHETIC

This is a distinction useful to keep in mind. Any scientific work mixes the two approaches. The ideographic considers the object of study as a unique phenomenon, whereas the nomothetic attempts to formulate generalizations. The ideographic approach, generally speaking, is inductive and descriptive. It relies more on intuition. The nomethetic tends to be deductive and systematic, it edicts laws.

From the perspective of a classification of knowledge, the hard sciences are on the nomothetic side, the social sciences on the ideographic. But the duality also exists within a single discipline, whether biochemistry or astrophysics. It is a question of temperament: some people are drawn more towards an ideographic style of research, or reporting, others by the nomothetic. The duality goes a long way in explaining the difficulty of the dialog between a scientist and a journal referee. Their worldviews differ entirely. When they argue a given result in a manuscript, truly they are expressing a fundamental opposition. It may well be that the author, having a more ideographic mindset, does not wish to go beyond the piece of datum to an interpretation. The referee, whose outlook is nomothetic, finds such behavior myopic at best and takes the author to task for not attempting to rise from the observation to the underlying law of nature. The opposite scenario can also occur, when a study takes a nomothetic view and attempts sweeping generalizations; while the referee enforces a positivistic attitude, sticking to the letter of the results and avoiding any overinterpretation of their meaning.

The distinction is also useful with respect to science communication. When a practitioner of the hard sciences addresses peers, the nomothetic style is more or less to be expected and compulsory. By conforming with the format of publications in scientific journals, one willingly adopts this approach, which in any case is the most familiar. When addressing the public, the message needs reformulating and reconstructing. It begs for a switch from the objective to the subjective, and the ideographic approach reasserts itself.

Hence we should remain aware of this very basic distinction. It operates in every field of research, it colors any piece of scientific writing, whomever it is addressed to.

REFERENCE: P.T. Manicas (1998) *Social science, history of philosophy.* In: *Routledge encyclopedia of philosophy.* Routledge, London.

the uncanny power of science

ILLUSTRATION

To combine image and text ought to be a winning combination. But is it truly? What are the conditions for success? Ought images and the text they illustrate be redundant, or just complementary to one another? Should they be alloyed into unity, or only allied in a form of parity? And what is it that makes an image interesting?

A lesson from history is that the answers to the above questions have changed markedly with time. As early as the thirteenth century, Western libraries carried manuscripts in which blocks of text were interspersed with technical diagrams. In the sixteenth century, printed reports from travellers were illustrated, not only with maps of the exotic lands which had been visited, but also with woodcut engravings showing the strange human types, the extraordinary plants and animals from the places visited. The Renaissance was also the time for the appearance of anatomical plates: they made possible diffusion of brand-new knowledge about the body, ushering-in a revolution in medical thought and in surgical practice. The seventeenth century witnessed an innovation of lasting value and consequences, the insertion in books of diagrams for concepts. Both Newton and Descartes, among quite a few other authors, made use of such illustrations. As for the Enlightenment, it created its own versions of diagrams as tools for thought. This was the time when the first bivariate plots and the first histograms were published (William Playfair, Jean-Henri Lambert).

From the mid-seventeenth century on, a key role of scientific illustration was to display the otherwise invisible. Robert Hooke was the pioneer, when he published in 1665 his *Micrographia*, i.e., an atlas of plates drawn from viewing a wide diversity of specimens through the microscope. John Dalton, about 1805, illustrated his atomic theory with schematic representations of atoms in molecules. Fox Talbot, one of the co-discoverers of photography, entitled *The Pencil of Nature* the portfolio of photographs which he published in 1844. The notion he advocated was that nature, through this novel means, was able to express herself and to draw her self-portrait – a remarkably naive statement. Eadward Muybrige, in 1887, with another album of photographs, collecting series of still pictures shot at short intervals, *Animal Locomotion*, was able to analyze movement and to describe accurately the strides taken by, e.g., a galloping horse. Just a few years later, Röntgen, with his famous X-ray picture of a hand, would depict in 1895 the ability of his new rays to penetrate matter and to image, as it were, its innards. A year later, Becquerel with his picture of a key would extend it to the mysterious rays emanating from salts of some heavy metals. All such pictures, from Hooke to Becquerel, demonstrated the uncanny power of science, for depicting reality in ways never seen before. Scientific images replaced the mythical images of yesteryear, the mermaids and the unicorns of legend, with novel sources of wonder.

But what are the roles of illustrations in scientific publications, if not multiple and diverse? I see six main functions. The first is documentary, to show that something is real, that it can serve as a piece of evidence. The second function of an image is methodological, when it depicts an experimental set-up, an apparatus. The third, non-identical with the first, is persuasive: an image shows that a model matches, more or less, the data. The fourth is also a little different; it has anteriority over the one just mentioned: heuristic use of a diagram, such as an histogram or a regression line, to try and make sense of the data. The fifth is to use an image as a short-hand for a description in words: a flow-chart plays such a role. Finally, an image often serves to communicate to the viewer the quality of the work done, in a rhetoric of excellence (one-upmanship and self-promotion at work).

As is well known, and as is to some extent an exaggerated difference, the two halves of the brain process differently visual and verbal material. We see things before we become able to name them.

Moreover, the visual has greater informational content (»*a good picture is worth a thousand words*«, this Chinese proverb can be trusted). Pattern recognition allows a reader of Chinese pictograms to give meaning to one among several thousand characters. In another form of visual literacy, a sequence of discrete images makes up successive frames in a story. Not to forget the hedonistic component, more easily fulfilled for most of us with images than with text.

What are, by contrast, the assets from a textual mode? It allows the convincing presentation of a linear argument, such as a series of logical deductions or inferences. It offers narrative continuity, where images – I am talking here about still images, not about those from a movie or a video – are by necessity discrete.

We are all familiar with iconic narratives. I am about to push such means as a highly effective tool for telling a story. In the presence of an illustrated text, as a rule people look at images first. The set of images in a scientific publication has a life of its own. Semi-autonomous, it tells a story of its own, at least its own version of the story. Respect for your readership dictates that you make this channel an effective summary of your message. If you devise it as self-contained, it won't duplicate your textual explanations, it will contribute its own, separate appeal and dimension. In science communication, the images often amount to evidence while the text amounts to rhetorical discourse, aimed at convincing its reader.

Not to forget: there is a third channel of communication, in quite a few science reports, the equation. It is a shorthand, with heuristic value. It consists, literally, of thought in action.

The text presents an argument, aims at convincing (I am repeating it for emphasis), and sums up at the end. Conversely, the image, akin to an exhibit in court, also pleases the eye.

The right brain hemisphere is predominantly non-verbal. It excels in the visual, it seeks to determine relationships between objects perceived or imagined in space.

The left hemisphere processes information sequentially, it has been described as the analytical half. It is more adept at processing verbal information. Verbally presented material, textual material equivalently, is encoded only in the verbal mental apparatus. Visually presented material is encoded in both the verbal and the iconic systems. This explains why, generally speaking, it ought to be easier to remember pictures.

Now to some recommendations. For a standard length paper, pick a dozen images. Make sure they are interesting, keeping in mind that each is, or ought to be, a story frame. Work on the captions, to make them concise and clear. If you provide your readership with a memorable set of images, you will have told the iconic part of your story.

What is an interesting image? In addition to the Chinese proverb, let me mention the rule of thirds in photography. Divide both sides of a picture, vertically and horizontally, into three segments. Thus, the image consists of nine juxtaposed rectangles. Place the center of interest at one of the four intersections, not at the center of the image.

It is useful to distinguish between literal representations that are intended to resemble the object they portray, and symbolic or schematic representations. Charts, graphs and diagrams occupy a middle ground. There is a continuum between realistic pictures, which resemble what they portay, and words, whose meaning is conventional.

Diagrams are subject to three rules, of specificity, selectivity and simplicity. They are specific. Any diagram carries a single concept. Overload it, and it sinks. Diagrams are selective. Any diagram selects from an infinite set of possibilities a single type. It is selective also with respect to the data chosen for depiction. And diagrams follow a principle of simplicity, or rather of simplification: a diagram operates like a special pair of glasses which, when trained upon a complex system, simplify into such an understanding, that a mere single glance suffices.

The main function of diagrams is heuristic, as a tool for thought. Charts such as maps, diagrams and graphs are effective in instruction, because they allow students to use alternative systems of logic. Learners can bring to bear their specific skills. Some people are able to recognize quickly geometric patterns. Others are less conversant with right-brain processing.

Devote great care and quite a bit of time to devising good captions. Ideally, the viewer is intrigued by an image and turns to the caption to better understand what is shown. A didactic mode is thus usually expected, one that is somewhat redundant with the image. Nevertheless, it has to remain brief. With choice words, describe this key scene in the iconic story.

And what happens if, conversely, one uncouples an image from the accompanying text? Turning them into entirely disjointed stimuli is often a nice trick. It gives the viewer a feeling of freedom: free to take pleasures from the images, free to gather information from the text. Entertainment and learning are separate but equal, they coexist on the page. The now defunct monthly magazine *The Sciences*, which was published by the New York Academy of Sciences, under the editorship of Peter Brown, made this its policy and, as it were, trademarks: articles on scientific topics were illustrated with reproductions of contemporary art pieces, which had little to do – quite a few were nonrepresentational – with the subject matter in the text. This was wonderful! I'll come back to this issue at the end of this segment.

Photographs, as illustrations, provide the documentary illusion: the reader can project into the laboratory, looks at the oscilloscope or through the eyepiece. Reproduction of the document with a picture is unavoidable with such evidence as an MRI scan of the brain, or images of atoms on a surface, as captured for instance by atomic force microscopy. Historical or archeological artifacts need likewise photographic documentation. The apparent objectivity of the medium gives the image the appearance, nay the illusion of reality.

However, several studies in the field of perception have concluded that subjects can recognize three-dimensional objects at least as well, if not better, when the edge lines (or contours) are drawn, by contrast with shaded or textured images. Thus, line drawings have to be privileged over photographs as an effective means for identification: we do not see photos of our politicians on the front page of newspapers, but caricatures which both simplify and exaggerate their characteristic facial features. The preference to be given to line drawings over photographs is consistent with Tufte's Principle of least Ink for effective graphic design.

Plotting two variables (x, y) against one another is, arguably, one of the main graphic tools for scientists. Whenever displayed in the ensuing paper, as a rule it is beset with obesity: too many data points. Use the smallest number which will support your argument. You will anyway state the total number of points in the text, when you provide a correlation coefficient for the regression (if linear). Such an image, known technically as a bivariate plot, is a great way to focus attention on the discrepancies. Truant data points may well be the most interesting to the investigator.

Scientists tend to pride themselves on linear regressions. What is good about this attitude is that it partakes of a quest for simplicity. However, data reduction does not stop with the finding of a correlation (remember: it does not imply causation). It entails building a model which will, not only explain the reason for the correlation, but also account for it qualitatively and quantitatively.

Pie charts display qualitative data. Histograms are a great tool to display and interpret accurate quantitative data: even minute differences in height jump to the eye.

Visual and verbal elements are in pronounced contrast. The former reveal, at a glance. The latter have a cumulative effect. The former are illuminating, the latter are argumentative. The former have the potential for being misleading, the latter can also be hijacked and this is the tall tale.

You have to exercise control, so that they do not conflict. Both kinds of evidence differ as to their strong points, as they differ in their capacity for delusion.

Such control is applied through the *offset* you choose to give an image with respect to the text it illustrates. If minimal, then the image has maximum documentary content, minimum symbolic content. Its chief role is didactic. At the other end of the scale, at maximal offset, the image has negligible documentary content and dominant symbolic content. Its chief role is oniric and subliminal, to massage the unconscious. In-between, going from weak offset to strong offset, the documentary content of an image goes from great to significant to minimal, while the symbolic content goes from light to attractive and to captivating. The corresponding intended roles of the image are, respectively, persuasive, the anecdotal, and enigmatic.

REFERENCE: E. R. Tufte (1983) *The visual display of quantitative information.* Graphics Press, Cheshire CT.

an organizing scheme 

INDEX

Which books don't need one? Perhaps a 35-page-long lyrical description of sunsets – St-Exupéry's *Little Prince* did not carry one. A romance, such as *The Bridges of Madison County*. A play.

Most other books are improved by having an index. Its name is transparent. Like the finger with the same name, this device points to something, the location of a particular item in a book. It does so by coupling the name of the item with the relevant page number. Experienced readers know how to use it, to look up the various appearances of a given name or topic. An index is a tool for information retrieval. It allows one to access the desired piece of information without having to browse or read the book. Or one may remember having read something in which case consulting the index is a shortcut to finding it again.

Since it is of such service to your readers, an index is an absolute *must* in a good book. The presence of an index is near-compulsory in any work of non-fiction. Furthermore, a good book deserves to close with a comprehensive, detailed index.

What do I mean by the phrase »a good book?« In science communication, books cover a wide range, from the journalistic to the scholarly. Only in the former is an index dispensable. It is a must in the latter.

There are specialists, within the publishing industry, whose job it is to compose indices. A professional indexer brings special skills and a kind of genius to the task. As an author, to interact with such a wizard is a most rewarding learning experience.

Often and in the less ambitious books, the author may be the most qualified person for compiling an index. My recommendation is to try and compose it yourself. Putting together the listing of names and topics in your book is no longer the ordeal it used to be when one had to write file cards and organize them alphabetically in a shoe box prior to typing a long list.

The personal computer has put an end to that. Software now allows you to highlight all the words that you designate for the index as you go along. After your very last editing, read the manuscript one last time for this sole purpose.

I have so far emphasized only the aspect of service to the readership. However, the index is also a tool for you, the author. It is a detailed analytical table of contents, magnified to the utmost detail. Hence, you can use it as an organizing scheme. It is not unknown for a conscientious author to compose the index for a book before writing a single line of the manuscript. By doing so, you will ensure that all the topics you wish to include will be covered, and all the authorities cited.

REFERENCE: T.M. Bernstein (1984) *The careful writer: a modern guide to English usage.* Atheneum, New York.

being distinctive 

INTRODUCTION

An overspecialized organism may face extinction. It lacks the flexibility to adapt to changing conditions. While overspecialization may not make science disappear, given social needs, it does it great harm. Nowhere is this more obvious than when perusing titles of papers or reading through their introduction.

Thus this will be my first recommendation. Write for the generalist, not for the specialized reader. This will compel you to take a bird's-eye view of your work: what did you achieve exactly? What were the bases for launching this study? Which will stand as your most robust conclusion? What are others in the field likely to pursue in direct filiation?

You need also to be reminded that you are competing for the attention of the readership in a flood of other printed material. Your prospective reader is leafing through the pages of a periodical on a computer screen, or negligently turning the pages of an actual journal, while giving full attention to information from the Web. Accordingly, your introduction needs to be much better than catchy. It has to be compelling.

How can you achieve it? By being distinctive. Identify what is unique, not so much with your research but with yourself, your personality, your voice. Colleagues are interested as much in your approach to science as they are in what your paper establishes.

Having expressed these essentials, delve into the more obvious parts. Writing an introduction is equivalent to throwing open your room, turning your private space into public space. A lab notebook becomes a blog. Your study becomes a cubicle in a crowded place, with onlookers behind your shoulder, watching your every sentence. Those are apt metaphors. They translate into these rules: (i) Spell out and make explicit what is a given to you. It is one of the functions of an introduction to tell why you did this particular study and what you hoped to achieve. (ii) Anchor your contribution into the scientific flow, i.e., refer to earlier work by others and to contemporary work by your competitors. You won't lose any credit by giving due recognition. You will gain in respect and you will only reduce future animosity.

I am not telling you to shy from waving your flag. Advertising is definitely involved in the writing of an introduction, but the softest of touches is needed. One-upmanship can be very destructive.

Don't omit telling the readers what they are about to find in the rest of your paper, if they keep at it. Your introduction provides an advance summary.

Now that you have heeded these instructions, you may have drafted three manuscript pages. You need to edit them. Condense your introduction into no more than two or three **paragraph**s, one page maximum *in toto*. In making it short and crisp, make sure that the writing is fluid and clear. Brevity is golden.

REFERENCE: J. Zobel (2004) *Writing for computer science*, 2nd edn, Springer, New York NY.

element of hypocrisy 

I OR WE?

For a paper with a single author should a scientist use the first person singular (I) or plural (we)? The latter, the more formal way of addressing one's readership, is also the more traditional. In support of this convention, one of the many lessons from the history and philosophy of science is to see the advancement of knowledge as collective and cumulative, not as an individual endeavor. Thus, in that view the role of any individual scientist should be played down.

Conversely, there is an obvious element of hypocrisy in the use of that style. It partakes of a whole set of other conventions, which when taken together make for a way of writing that is rather leaden and for the resulting poor readability. I shall only mention here the systematic use of the passive voice, the one-upmanship manifested in statements of »we are pleased to report« type, and an overall legalistic style in which any assertion is accompanied by its qualifications.

Why not write in the first person? It has the double asset of being more genuine and more direct. It forces the author of the paper to present his or her argument in the form of a narrative – which is good since it is likely to induce greater interest from any readership.

Indeed, if the history of science can guide us in deciding between these two options, we ought to remember that scientific publications originated with private correspondence. Letters by philosophers-scientists reported their findings, and they were gradually disseminated to a whole network of acquaintances. Indeed, if we want to follow tradition, then we ought to go back to such a conversational style as was used in the seventeenth and eighteenth centuries before the advent of scientific societies and the first professional journals.

If we do so, this may be perceived in the community as a sign of arrogance. »Who does this upstart think he is?«, a few of our colleagues may grumble. And they may be in the right. It takes high quality indeed for any work of science to be presented with pride, with the pride attached to its lasting value. Is such a self-confidence justified?

Thus, the choice between »I« or »we« boils down to a question of whether self-assurance is warranted on your part. There is a pro, but there is also a con. Con: you know very well that any conclusions are tentative, that once published they will undergo critical scrutiny from the scientific community and that, in the long run, they are more than likely to become superseded. Pro: you deem your results important, you have checked them with the utmost care, and you stand by them.

REFERENCE: R. Barrass (2002) *Scientists must write.* 2nd edn, Routledge, New York.

a lilt to your prose 

IRONY

To write a piece in an ironical mode or to sprinkle bits of irony over a paper is not reprehensible. To the contrary, it gives a lilt to your prose, it makes it more buoyant. Moreover, if you are writing a review of any kind – be it of a research proposal, a **referee report**, a **review article**, a summary of a study from another research group, … – then an ironic tone becomes near-compulsory.

Irony belongs to the same family as wit and sarcasm. It is a subdued form of humor. Without your being outright funny, which almost always is unacceptable to editors of scientific journals, a touch of irony, a few self-deflating comments will pull in readers, and put them on your side, while lightening your piece.

After all, playfulness is essential to science. It colors everyday life in the laboratory. It can find its way into publications through the private language of the laboratory turning into an accepted shorthand expression in the literature.

For instance, certain DNA-protein conjugates are devised with the help of chimeric molecules comprised of a protein head covalently attached to an oligonucleotide tail. Their nickname, »tadpoles«, has now migrated from the Molecular Sciences Institute in Berkeley, California where it was coined, to the open literature, such as the magazine *Nature*. In echoing it there, Garry P. Nolan writes (ironically):

> *»Brent and colleagues applied (…) molecular gymnastics to create a novel detection reagent, which they quixotically term ›tadpoles‹ for their apparent shape and the series of metamorphosis steps they must endure until their final maturation and application.«*[1]

The words »gymnastics« and »quixotically« in this sentence bear the color of irony.

If irony needs to be distinguished from wit which it closely resembles, it also needs to be carefully distanced from sarcasm. Sarcasm in print is an unacceptable form of aggressivity, even though competition between research groups sharing similar goals fosters it. Such expletives should stay within the walls of your laboratory. You can project your competitiveness in other ways, by spicing-up your research reports with the salt of irony. Thus, you will mute your criticisms of what you may consider

– whether rightfully so or not is immaterial – shoddy work, overinterpretation of the data and intolerable self-aggrandizement.

No one will take real issue with small, ironical jabs, especially if some are turned against yourself in a wry, self-deprecatory manner. Any scientific communication needs to be tinged with doubt, if anything as a portent of the critical reception the scientific community will greet it with.

Irony will make your voice stand out. This is also a necessity. You want to appear as an original and a creative scientist, rather than just yet another anonymous contributor to the public discourse. Your approach should be distinctive: mark it with irony.

Irony is defined by dictionaries as a contrarian trait. Even though this is true, it is nevertheless too strong. I prefer to consider irony as a verbal tease. It is a way of neither taking oneself nor one's antagonists with utter seriousness.

Life is a comedy, at least for part of the time. Scientific life is no exception. To write ironically is to heed that fact.

REFERENCES: G. P. Nolan (2005) *Tadpoles by the tail.* Nature Methods 2:11–12; the primary publication which Nolan comments upon is: I. Burbulis, K. Yamaguchi, A. Gordon, R. Carlson, and R. Brent (2005) *Using protein-DNA chimeras to detect and count small numbers of molecules.* Nature Methods 2:31–37.

riding the wave 

NEOLOGISMS AND EPONYMY

Why do some scientists come up with new words, which they impose on their readers and listeners? Do they enrich or pollute the language? The good reason, which applies to perhaps a tenth such additions to the lexicon, is new concepts that need new words. The real reason is advertising of one's work, one-upmanship. The habit thrives on the need for novelty.

Such a need, respectable in itself, has become as unconscionable, as rude as scratching one's crotch in public. So many of the **opening paragraphs** in a publication carry phrases which overuse has made meaningless such as »for the first time«, »we have recently disclosed«, »the use of X has emerged as a new and important field«, »we have recently reported«, etc. Jaded as we have become, such wordings get under our skin like some of the most blatant attempts at manipulation in TV commercials, which are not even funny.

Let us look at an actual example. It explains the constant need of language to incorporate new terminology. It also shows scientists riding the wave of these additions. The word »surfactant« has entered common language. A surfactant molecule reduces surface tension, generally in an aqueous medium. A sub-class is known as »gemini surfactants«. A surfactant molecule comprises a head group and a tail, made of a hydrocarbon chain. Gemini surfactants have two head groups and two tails. The two moieties are bridged by a spacer, which connects either the head groups or the tails. The term, gemini surfactant, was coined in 1991 by Fredric M. Menger, who in 2000 also published a review article on these compounds.

The new term is a shorthand, which explains its entering general use. It might have been different. Perhaps, Dr. Menger also considered other names such as, to quote only a few, binary surfactant, tandem surfactant, joined surfactants, twin surfactants, dual surfactant, duplicate surfactant or even catamaran surfactant, which is exactly what it looks like, schematically.

»Gemini« was an inspired epithet, everyone knows this sign of the zodiac. It has a strong symbolical meaning, drawing from astrology, it is a source of amusement upon transfer to the scientific sphere. Moreover, when Dr. Menger introduced the new term, everyone in the field knew him as the progenitor. Accordingly, to refer in a publication to gemini surfactants was an implicit homage to Dr. Menger, which cannot have displeased him.

However, any metaphor is loaded and the excess baggage can become a nuisance. If the Gemini sign is well-known, most graphical representations show a head-to-tail disposition of the twins. Gemini surfactants, conversely, show a head-to-head arrangement of their two moieties. Which is why I referred above to »catamaran surfactants«.

Dr. Menger was no stranger to introducing new terminology in the literature, he was familiar with this form of advertising. Some of his publications in the nineties refer to »chemical collectivism«, or to »chemically-induced birthing and foraging in vesicle systems«.

The take-home lesson? Better to refrain from coining a new term. It will make you look brash and condescending. Do it only if your work warrants it and is truly superior.

REFERENCE: I. Verdaguer (1996) *Making sense of neologisms.* Forum 34(3):98.

for critical asides and irony 

NOTES

Information is characterized as either essential or not, as central or peripheral – to an argument for instance. Information that is both essential and peripheral belongs in the notes, not in the body of a text. What about segments of writing which are uninformative but somehow essential? They may be part of the logic or of the rhetoric. How does one tell, for instance, if a statement belongs in the text or in a note? Anything that gets in the way of the main argument, anything of interest, be it of marginal interest, any hindrance to fluidity of a read, has to be tucked away in the notes.

Notes are important, in the same way witnesses are important in a trial. They testify to your not having made up your story. Whereas a text displays a polished prose, the notes manifold gathers all the beams, struts and buttresses which the text needs as supports. They serve as a counterpoint, for critical asides and irony, to qualify an assertion, to provide bibliographical information about the source of a statement. Notes are not meant for the general reader, for someone who has picked up the book at random or having browed through it and been hooked. Notes are addressed to the scholar, to the expert.

How should you write them, then? In a note, information is at its highest density, concise in expression and compressed. Footnote or endnote? This depends on the house style for that publisher, for this journal. I prefer personally footnotes, which a quick glance at the bottom of the page reveals. To go back and forth between the page one is reading and the back of the book, or the end of the article, is more work. It is too much work, again speaking personally.

The British share with the German a lovely tradition, the bibliographic essay. It serves as an alternative to footnotes or endnotes. The British put it at the end of the book, the German put it in front. The characteristic feature is a narrative of its own. A bibliographic essay makes for separate reading, no less enjoyable than with the text proper, just different. In such a piece, the author shares with interested readers a critical review of the literature. Each reference is evaluated, not only for its pertinence, but for its quality too.

REFERENCE: A. Grafton (1997) *The footnote.* Harvard University Press, Cambridge MA.

a bait 

OPENING PARAGRAPH

To quite a few of us, composing the opening paragraph is an ordeal. We find ourselves tongue-tied. If we put something down, it stinks. We feel humiliated. We can't write. A double-expresso later, we are at the same point. We feel disappointed and highly frustrated.

What should we do? How to go? Are they guidelines or recipes? What are the rules?

Here is a suggestion for starters. Identify three elements in your work. These might be the method of study, the material observed and your main conclusion. Associate a word or a phrase to each such element. Now compose a full sentence which will include each of the three words or phrases. This provides you with three sentences. Set these in the most logical order. Now, you need some sort of a transition between sentences 1 and 2, 2 and 3. After you have thus glued them together, you are looking at the first draft of your opening paragraph. But you can improve on it markedly. Shorten it. Make it more informal, more natural and congenial.

Now, to another way of attacking your task. The opening paragraph is your hook, as they call it in journalism schools. It is the device with which to hook the interest, the attention too of the readership. Can you think of a fun manner to start your paper, something that will be unexpected, something which you will have fun presenting to the readers?

A third way, rather standard, is to devote the first paragraph to an announcement. You tell readers what is to follow. The opening paragraph will give away the gist of the paper. It tells people: »Here is what I am about to tell you. My story will have the following parts. First, I shall cover the background to this investigation. Second, I will be telling you of our methodology. Third, you will be shown our results. Fourth, will come the interpretation we submit for our results, whose meaning as you will see is XYZ«.

Let us look at actual cases. A recent paper (autumn 2004) starts with the sentence »*Speculating about the future of science seems to be genetically encoded in scientists.*« Excellent! It intrigues. Whether the assertion is true or not is irrelevant, the conjecture is witty. One is impelled to read on. Another opening sentence from the same journal reads »*The drive to shrink electronic devices to the nano-level, has, in recent years, led to the design and investigation of molecular-scale components endowed with sensing, switching, logic, and information storage functions.*« Too long. Breaking it into separate sentences improves it: »Electronic components have sensing, switching, logic and information storage functions. In recent years, one witnesses a drive for their miniaturization down to the nano-level, the molecular scale.«

I cannot overstress the importance of your opening lines. To invest a couple of hours just in writing your opening paragraph is well justified. The hook has to be made attractive. Find a bait, so that you make a big catch!

REFERENCES QUOTED: G.M. Whitesides (2004) *Angew chem int edn engl* 43:3632–3641; M. Ruben et al (2004) *Angew chem int edn engl* 43:3644–3662.

articulate your viewpoint 

ORGANIZING YOUR MATERIAL

The two obvious steps are the organization of the documentation and of the story you tell. Your documentary material should consist of some research report or results, evidence in favor or against its conclusions, some kind of a logical argument, graphs of the data, and opinions from a number of people who may have been approached in person, on the phone or just from reading the literature. After going through the whole material, make a list of points – typically no fewer than a dozen and no more than two dozen, the numbers give you an idea of how coarse-grained your organizational structure can remain.

Number these various points in sequence. The order you pick does not matter, as long as you are consistent. It can be the logical order, that of steps in a demonstration or in a persuasive argument. It can be the chronological. It can be any old order. Be aware though that your organizing principle will affect the narrative you compose.

Your first need is indeed to plan ahead and to choose how to segment your text or presentation. You will be ordering your documentation as a function of such a plan. For instance, Richard Dawkins in his little book *The View From Mount Improbable*, [2] in which he considers the eye as an organ optimized through Darwinian evolution, first introduces cup eyes from the animal kingdom; he then goes on to pinhole eyes; thirdly, he presents lenses-endowed eyes. His obvious organizing scheme is from the simplest to the more involved and complex, a move generally to be recommended.

In organizing any intellectual piece of work, I personally find most helpful the compiling of a list. Anglo-Saxon culture has an admirable trait, reasoning by points: first listing the various points – i.e., aspects, angles, topics, …in a subject – and setting them in sequence only when a sufficient number has been collected.

But what kind of a sequence? For lack of a better organizing scheme, one well-suited to your material, you may want to opt either for the *logical* or for the *chronological* order. These almost always work fine.

It will often help to write out the opening paragraph of your article, chapter, …, at the same time as you draft a list of the topics to be covered. These initial sentences will haul behind them those which are to follow. They will thus order gradually, one by one, the whole gamut of items on your list.

Early planning is indeed a necessity. It will make you aware of the gaps which remain to be filled in the documentation to be gathered, prior to any write-up.

Let us assume this first step has been satisfactorily fulfilled. You now need to organize the argument, so that it be logical and proceeds in nice steps. As soon as you articulate your viewpoint, try to think of other takes on your material. Do not utter the *pros* without considering also the *cons*, all the more so that you have refutations at the ready.

As an example of a nicely-flowing narrative stemming from a transparent logical organization of its material, Atul Gawande's »The Bell Curve« essay in *The New Yorker* has these parts: 1. a case history of a debilitating illness, cystic fibrosis; 2. statement of the problem, hospitals in the United States differ markedly in their performance towards such patients; 3. a description of the performance-based grading of hospitals, a rapidly developing practice at present (at long last?); 4. a portrait of a militant for such grading; 5. a description of a standard program of cystic fibrosis care, as performed in a specialized ward in Cincinatti; 6. by contrast, description of the more aggressive, and much more successful approach taken in a Minneapolis hospital; 7. portrait of the Minneapolis innovator; 8. in conclusion, besides being graded on their results, doctors need to constantly strive for excellence.

My recommendation then is to compile an analytical table of contents prior to putting pen to paper. You only need to flesh it out afterwards, when you tell the story. An analytical table of contents is what can still be found in a number of books, of nonfiction especially. It is what we enjoyed (or infuriated us) in those old-fashioned British novels of the eighteenth and the nineteenth century, with very detailed chapter headings, such as (I am making them up): »23. In which the General does not go to war but nevertheless finds his match.« or »17. Our hero goes to college, becomes sick, quickly recovers and is elected to the debating team.«

An important case is the Discussion part in a primary publication. Needless to say, you will be presenting your interpretation of the results. Remain aware there are other viewpoints too. Thus, you need to play the Devil's Advocate. List the objections to your interpretation. List them from a to Z, trying to be as objective as possible, i.e., having steeped yourself in the firmest of doubts, wearing the robes of St. Thomas. Then proceed to review those counterarguments one by one, settling each on its merits. Your having done so will immunize your paper against the harshest among possible criticisms. You will be able, then, to close your paper by pointing to directions for future study.

Once your text is in working order, turn to the illustrations and do the same. They need to tell their own version of the same story. Order them in sequence so that each image frames an episode in your story. Make sure that the train of illustrations tells the same tale as the text does, although in its independent way.

Such preparatory work, the detailed overall planning of an entire paper or even of a whole book, requires a great deal of thought. But it is worthwhile, a good investment.

There is a yet better option. I have tried it myself a few times. It applies only to the writing of a book, though. As your very first task, put together the **index**. After you are done listing all the entries which you think ought to be part of the index, work backwards. From the list of topics thus defined, cluster them into sections and chapters. Finally, do the writing. The chief merit of this admittedly unusual procedure is to prevent omissions. Thus, consider the index as both programmatic and recapitulative.

Now and only now you should start writing your story. You know how to organize it, from having heard or read so many stories: 1. set the stage; 2. present the characters, both the »good« guys and the »bad« guys; 3. present the task in front of the good guys; 4. go through the obstacles they face, including the resistance from the bad guys; 5. if need be, have a shoot-out scene; 6. tell how ultimately the good guys prevailed; 7. if you are writing an essay of moderate length, neither a short newspaper piece nor a book, you may want to end, in short story manner, with an unexpected twist to the story.

Last but not least, *where* should such ordering take place? So that it be not disperse, and that you won't have to look in too many places for this roadmap, I suggest that you write it down in a commonplace book, or in the equivalent within your computer.

REFERENCE: A. Gawande (2004) *The bell curve*. The New Yorker, December 6: 82–91.

informative words 

PARAGRAPH

»After all these somewhat abstract mathematical generalities I am now going to show you a few pictures of surface ornaments with double infinite rapport. You find them on wall papers, carpets, tiled floors, parquets, all sorts of dress material, especially prints, and so forth. Once one's eyes are opened, one will be surprised by the numerous symmetric patterns which surround us in our daily lives. The greatest masters of the geometric art of ornament were the Arabs. The wealth of stucco ornaments decorating the walls of such buildings as the Alhambra in Granada is simply overwhelming.«

(from Hermann Weyl (1952) *Symmetry*.
Princeton University Press, Princeton, NJ, p. 109)

A close read of this excerpt from a classic of twentieth-century science makes us aware of the ingredients in a successful paragraph. It is both rich and easy to read, because Weyl chooses highly informative words. »Stucco« is one such word. The phrase, surface ornaments, is meant to contrast with »mathematical generalities«. It does so subtly. Instead of pitting the adjective »concrete« against »abstract«, Weyl chose the word »ornaments«. This noun refers unambiguously to crafted objects, i.e., to actual objects as opposed to mathematical idealizations. Moreover, this noun announces the sentence which follows and gives examples of such ornaments.

Weyl is a master of the subtle contrast. Another example, from the same paragraph, is »prints« versus »dress material«. These are not only apt words, they are evocative too. The second term, prints, indeed conveys an image of surface ornaments, often repetitive in design, consisting of multiple adjacent copies of a pattern. The key word here is the adverb »especially«. It draws attention to the first term, »dress material«, reminding us that woven fabrics in general indeed show repeating patterns.

The next phrase, »Once our eyes are opened«, is quite effective. It is truly the pivot of this paragraph. Having made us attentive, which was the purpose of the preceding words, the writer reinforces such heightened perception by making us be aware of it. What he is really doing is encouraging us to take stock of the wide diversity of symmetric patterns to be encountered in daily life.

The last two sentences serve to substantiate the claim, to exemplify it historically (the heyday of Arabic culture) and geographically (Andalusia), and to draw our attention, again with the gentlest of nudges, to such decoration occasionally rising to the level of high art.

This paragraph shows that science communication can forego mathematical equations, it can dispense with technical jargon. The paragraph is devoted to a single idea: we see symmetric patterns all the time. We ought to take better notice of them.

Furthermore, the paragraph has a point of departure (math) and a point of arrival (art history). In between, it flows very nicely. The »overwhelming wealth« in the last **sentence** neatly pulls together all the elements the paragraph has launched. The phrase is even generous: the writer anticipates and forgives our having been inattentive to the wealth of symmetric patterns around us.

REFERENCE: J. M. Williams (1994) *Style: ten lessons in clarity and grace.* HarperCollins, New York.

this elusive quality 

SCHOLARSHIP

A scholar is someone who knows a specific field in depth. Such a definition clearly is not sufficient. We have to elaborate upon it. Scholarship is what prevents the phrase »narrow specialization« from redundancy. Scholarship identifies with broad specialization. Yes, this is something of an oxymoron. There seems to be a contradiction between being specialized, i.e., positioning oneself at the extreme point of knowledge in a given area, and for this sharp point to somehow have diffused into a whole arc. Which explains why few scientists are able to endow themselves with this elusive quality of scholarship. We shall see what makes it desirable.

But we need first to gain a better understanding of what scholarship consists of. One might also term it an old-fashioned quality. The word »scholar« is a relative of »scholasticism«, the school of thought which the New Science of Galileo, Descartes and Newton had to upset, in order to usher in modern, experimental science.

Scholasticism was dogmatic, steeped in the writings of Aristotle and the teachings of the Church. Such a connexion had left scholarship with a somewhat negative connotation: a scholar, in such a jaundiced view, is a person holding on to past knowledge, timorous in the face of novelty, someone who is steeped in, even brainwashed by, the old paradigms.

Accordingly, present-day scientists, new to the pursuit, perhaps unimaginative or not knowing any better, deem scholarship useless and irrelevant to the practice of science. But let us return to the definition of scholarship as »knowledge in-depth«. This equates with **erudition** in the attendant literature. A scholar is someone who is not only familiar with the literature of the field, but who somehow transcends such familiarity. A scholar's brain can immediately give an accurate reference for X, to connect Y and Z and, in so doing, makes one despair of ever emulating that ability.

The counterargument here is: is it not the case that the **Internet** now allows one to dispense with scholarship, since it offers, on any topic, at the instigation of a search engine such as Google, the whole world of scholarship? Indeed, there is even a subset of this leading search engine known as Google Scholar. To answer the charge is easy: in order to look for something, one has to know in advance what one is looking for.

How then to acquire scholarship? This ability goes with disinterestedness. It is the product of intellectual curiosity. If you do not have the inclination to poke around, to acquire seemingly useless knowledge, to put your work into a wider frame, such myopia will leave you blind with respect to scholarship. The only way in which to gain it is to be adventuresome in your scientific pursuits. For instance, browsing the pages of a journal, while looking up some precise information from a paper in that journal, is one of the efficient ways in which to build up one's scholarship. If this is the way toward acquiring it, how then does one teach scholarship? By the virtue of example, first and foremost.

Scholarship, however, is neither necessary nor sufficient for the advancement of knowledge. It is not necessary: there are discoveries made in a flash such as, for instance, that of the PCR procedure for copying strands of DNA. Cary Mullis did not need to be a scholar in order to invent it. That his breakthrough qualifies more as an invention than as a discovery is a valid point.

Neither is scholarship sufficient to ensure that a contribution be of value. There is the case, all too frequent, of the fine scientist who applies skillfully to the problem at hand all the relevant procedures, who furthermore is well-versed in the literature of the field, and who nevertheless fails to make a deep impact – for lack of vision.

Enough said. To discuss scholarship is to uncover paradoxes and apparent contradictions, one after the other. I shall leave you with the central paradox of scholarship in science: innovation thrives on knowledge of things past.

REFERENCE: C. E. Glassick, M. Taylor Huber, G. I. Maeroff (2000) *Scholarship assessed: an evaluation of the professoriate.* Jossey-Bass Publishers-Wiley, Hoboken NJ.

felicitous choice 

SENTENCES

A sentence is the fundamental unit in any piece of writing. As such, it has to work well. You will need to revise your first draft. You will need to keep working at each sentence to perfect it until it fulfills its role.

To become an invisible and efficient piece of textual machinery, a sentence has to be logical, precise, well-built. It should sound nice, read aloud. It ought to be concise. This provides us with five criteria, logic, precision, simplicity of structure, euphony and brevity.

But let us consider a couple of examples. Here is exhibit A. This actual published sentence reads:

»Modern culture trains us to expect instant answers to every question, but reading encourages slow-fuse inquiry.«

Let us apply the above criteria, starting with the need for statements to be logical. Do we expect answers, in the plural, to every question, or just a single answer to every question? The latter. Do we expect an instant answer to every question? Unlikely. This is an exaggeration. Reading, if considered in a questions-answers mode, does not encourage *inquiry*. Rather, it may provide *answers* to one's questions. »Slow-fuse inquiry« is a non-starter. What is clearly meant is »slow-fuse answers«, which makes sense, not »slow-fuse questions«, which is what »slow-fuse inquiry« amounts to. Is not reading also an integral part of modern culture? A culture does not »train«. A culture molds, induces, incites, encourages, breeds conformity or dissent, …

Application of criterion number one thus shows Exhibit A to be marred by logical flaws, even absurdities. To remedy them is the first task. A first revised version thus might read:

»We may expect instant answers to questions. Yet, reading only yields slow-fuse understanding.«

I have broken the original sentence into two, simplified the structure, and removed the logical inconsistencies. In addition, a slow-fuse is a delaying device. Hence, we should strengthen the sentence by replacing a noun – a nominal phrase actually – with a verb, thus:

*»We may expect instant answers to questions.
But reading delays the fulfilment.«*

This second revised version suffers from the second sentence lacking a strong coupling with the first. A better, third revision is accordingly:

*»We may expect instant answers to questions.
But reading delays them.«*

I won't belabor this example further. I shall only note that the very first criterion, the need for a sentence to be logical, has proven its worth. Moreover, in so doing, we have also satisfied the fifth criterion, that for concision. We have reduced Exhibit A from its original 96 characters to 58 characters, a gain by 40%. Our editorial work has produced something clearer than the original prose, something now readily understood (if rather mundane).

Exhibit B is a sentence from the pen of Richard Dawkins, of *Selfish Gene* fame. This biologist is taking-up the question of organisms evolving eyes, using images of objects the right way up, rather than upside down. Two successive sentences of his read:

»But the physical difficulties of turning the rays round are formidable. Amazingly, not only has the problem been solved in evolution, it has been solved in at least three independent ways: using fancy lenses, using fancy mirrors, and using fancy neural circuitry.«

The first sentence is graced by felicitous choices: (i) »turning the rays around« is likely to have been the first draft. The correction to »round« instead of »around« is good, euphonically, it brings in the alliteration rays-round; (ii) it is written in problem-solving approach, from stating that the »difficulties are formidable«; (iii) it uses only plain language, skirting the technical term (inversion) with the paraphrase »turning the rays round.«

The second sentence flows directly from the first: notice how »amazingly« answers »formidable«. Notice how »amazingly« is echoed in the triple use of the adjective »fancy«. The mood in these two sentences, their color (yes, sentences can be endowed with color), is wonderment at the beauties of nature, evolution in this case. Furthermore, this second sentence in Exhibit B makes us want to read on. It sketches what will follow. Obviously, Dawkins will next tell us what the three solutions consist of, using lenses, using mirrors or using neural circuitry.

With these two examples, I meant to emphasize how a sentence has to be crafted with care, if it is to carry meaning effectively.

REFERENCE: W. Zissner (1990) *On writing well: an informal guide to writing non-fiction.* 4th edn, HarperCollins, New York.

the invisible, global stockmarket 

SIGNERS

Why does authorship of scientific papers carry such a massive weight? A reason is moral responsibility for one's offspring. Another is the quasi-heroic feat inherent in almost any piece of scientific research. On the average, we are told, a scientist publishes *a single* paper over his lifetime, printing the gist in his Ph.D. dissertation. If this is the norm, then ownership of a whole list of publications becomes all the more respectable.

But is it really? In the idealistic enthusiasm which accompanied the student revolution of 1968, there were proposals to entirely abolish such signs of private property. Some came out in the open literature. Several papers originated anonymously with »Groupe des cristaux liquides d'Orsay.«

To publish is comparable to issuing stock. Your name henceforth becomes valuable in the invisible, global stockmarket in which reputations, good or bad, are made. The stockbrokers are your entire community of peers. Is your name a blue-chip stock, a sound investment, or is it rather a junk-bond, not worth the paper it is printed on? There are many shades in-between these polar opposites. The analogy, moreover, extends to speculative bubbles. For instance, the names of Stanley Pons and Martin Fleischmann, associated with the cold fusion debacle do not ride very high in present public trading.

A related question is, who should be included amongst the authors of a scientific publication? As a rule it results from teamwork. But who qualifies as a member of the team? Obviously not the commercial supplier of chemicals, even though those were needed for the work. But where to draw the line? Be very restrictive in this matter. The criterion I submit, and deem excellent, is for every single person whose name appears on a paper to be able to both present and defend the entire piece of work. If this exclusionary principle were to apply, about 30% of the current authors would see their names deleted.

The final question is that of the ordering of the names of the authors, if there is more than one. A common usage is to set the name of the group leader first, followed by the lieutenant, and so on, down to the simple soldiers. Such a military ranking, even though frequent in current practice, reflects a single reality, seniority, rather than the actual commitment various people made to the work. When the scientist who has carried out the brunt of the work is put last, it legitimately rankles.

Another custom is to set your name last, preceded by the list of those of your coworkers in decreasing order of importance to the work. This improves upon the preceding habit. Nevertheless, anyone in the know – ultimately, everyone along the grapevine – will refer to this publication by attributing it to the senior author, anyway (which is helped by the asterisk denoting the person one should write to, for reprints, additional information, or anything else). Thus, such a listing may create confusion, to no one's advantage.

Personally, I prefer the alphabetical order of names. This is a British tradition and it has much to commend itself. It is currently in disuse, which is unfortunate. Returning to it would solve many problems.

REFERENCE: S. Bachrach, R. S. Berry, M. Blume, T. v. Foerster, A. Fowler, P. Ginsparg, S. Heller, N. Kestner, A. Odlyzko, A. Okerson, R. Wigington, A. Moffat (1998) *Intellectual property: who should own scientific papers?* Science 281(5382):1459–1460.

do not forget to breathe 

SPEECH DELIVERY

Practice makes perfect. Nowhere is this more true than in the intricate art of public speech. Experience makes people good at it. To mention professions for which the vocal dimension is foremost, actors and teachers usually excel at it.

Rehearse often your presentation, not only isolated segments but also at least one nonstop run-through.

There is a great deal you can do to improve your own performance.

First, listen to yourself. Watch a video of your speaking or even reading aloud. You will be your own best critic. Note the defects and the annoying mannerisms. Return in front of the camera for an improved delivery.

There are pre-requisites to good speech. Set your body properly. Be dressed comfortably. Leave your neck free. Your chest and your feet should likewise feel unconstrained and of course painless. Discover which resting positions work best for you. Arms closed across your chest? One hand in a pocket, or posed on the lectern? Make a repertory of at least three such stands between which to alternate. Crucial is the anchoring of your whole body onto a »support:« for instance, a pencil held in your hand or between your hands can serve as such.

You do not have to remain static, to the contrary. Watch performing artists on a stage. They keep moving, don't they? Follow suit. Find which motions come to you naturally. Obey the urge and espouse their rythm.

You are almost ready for **taking the floor**. Breathe deeply, make several conscious inspirations and expirations. They will help oxygenate your brain adequately, and keep you alert. Start speaking. Memorize and perhaps even rehearse, to someone, or to your mirror, your first lines. They will thus come easily to you.

Try not to read your talk. It will sound much better if seemingly spontaneous. You can help with cue cards to look at from time to time. Your audience will respect your need for an occasional pause, and for taking a breath – literally. They need it too.

While speaking, do not forget to breathe. Make your **sentences** short enough that you can utter each in a breath. Pace yourself. A public speaker is like a horse running on a track: once a pace has been chosen, to change it is tricky. Make sure to talk conversationally, neither rushing through the words nor being agonizingly slow. Study your written text sentence by sentence and mark it. Your breathing needs being brought in harmony with the punctuation. Your delivery will have to convey it by itself.

Volume? You want to be heard throughout the room. Avoid being too loud. Exercises will teach you to project your voice. For instance, say in succession the same word or a sentence, such as »can you hear me?« in order to be heard at distances of one, five and twenty meters. During your talk, make sure to vary occasionally the volume of your delivery, to avoid monotony. For emphasis, you may want to say some of the things in a tone of confidence, which goes with a softer voice – a trick which every teacher uses on a rowdy room needing to be silenced and to become attentive.

Likewise, give different colors to your voice. An extremely useful exercise is to repeat a single word or sentence, such as »condensation of droplets«, in various tones, those of assertion, logic, anger, sharing a secret, being lewd, etc. Such exercises are best performed with a tutor. At times, you may want to emphasize a single word: pronounce it slowly, over-articulate and detach syllables from one another: con-den-sa-tion.

Pitch? It is important that your voice be pitched at its natural, strain-free level. Learn how by reading a text aloud. If you are a beginner, do such an exercise five minutes twice daily. Singing, if you are so gifted, will help you greatly. Being tutored by a professional works wonders. Watching your audience – your audience, not the projection screen – , scanning their faces and making eye contact, these will help you to find and maintain your pitch.

In so doing, find your own voice. This is the basic secret for seducing your audience. If you make people feel there is a genuine person talking to them and sharing knowledge with them, they will bear with you to an amazing extent.

Some people are incapable of it. They speak in an affected, unnatural voice. This unfortunate personality trait probably dates back to teenage years, or even earlier. Their assumed voice is part of their mask, of their public persona. Any audience detests such phonies. The very word »phony« says it all. If you are unfortunate enough to be such a person, you will have to undergo a major change and come out of your hiding place. It may well be that only a psychotherapy will do.

Such a pathology serves to show how fundamental a voice is, to interpersonal relationships and thereby to being an efficient communicator.

REFERENCE: S.L. Tubbs and S. Moss (1999) *Human communication.* 8th edn, McGraw-Hill, New York.

a maximum of 20 characters 

SUBTITLES

These pertain to an article for a scientific journal, for a magazine or, sometimes, in a book chapter. They serve two main functions, to fence off segments in your text and to announce the contents of each such segment.

The first function is typographic. However, you won't have a choice. You will be given the house style favored by the publishers. Traffic signs are visually clean, easy to see and to understand. The format for the subtitles is usually likewise clearcut.

The second function is communication. While it is not your prerogative as author, you will be able to discuss and negotiate with your editor the wording of the subtitles in your piece. Make your subtitles neutral – do not try at being funny or cute – , informative and concise. I recommend a maximum of 20 characters for each subtitle.

REFERENCE: J. J. Gartland (1993) *Medical writing and communicating*, University Publishing Group, Frederick MD.

to grab the attention 

TAKING THE FLOOR

Do it. The expression is to be taken literally. You have been given a chunk of space-time. Your first task is to stake a firm claim over it. Take possession of your territory, of your whole territory, left and right, behind you until the projection screen, in front of you until at least the first row of seats. Be mobile.

A silent walk back and forth, during which you appropriate your space prior to launching into your lecture, helps listeners to focus on you. They will stop conversing. Prior to taking the floor, get prepared. Make a series of deep breaths, inhaling and exhaling slowly. If possible, get it done outdoors by yourself with clean, fresh air. You need to be alert, ventilation oxygenates your brain.

You are now being introduced. It is your single opportunity for surveying your audience. Feel free to address a wink or a smile to people you know. Take register of the various attitudes here and there, you will have to respond to them: expectancy, curiosity, goodwill, or even possibly, but rarely, downright hostility.

After you have been introduced, improvise the first two or three sentences. You will feel, and sound natural. This will help you to slide into your talk proper. Because you need to grab the attention of the audience, your opening lines should be terse, well-built, elegant and distinguished. Write them down in advance and memorize them. The same holds for your closing statement. Keep in mind the Golden Rule of public speech: start by telling people what your story consists of, then tell it, and finish by summarizing what you have said.

Try to feel at ease. Dress comfortably. Have a drink and your throat lozenges within reach on the lectern, or on a nearby table. Stand straight. Be at the same time poised and relaxed. And breathe! Do not forget to breathe at regular intervals. Believe me, it is not fun to watch someone losing voice, for having forgotten to breathe, in the excitement of **speech delivery**.

Each of your air intakes is the engine propelling half-a-dozen sentences. You need a brace, a support, an anchor for your body. You might place a finger or a hand on the lectern (beside rather than behind it). You may hold your pen, your glasses ... Your hands are being watched, too. Use them. Use them a lot, for expressiveness.

Be not afraid of silence. Underline major words, major concepts, major points by a short, silent follow-up, to let them sink in. For additional emphasis, repeat important or difficult words occasionally. Repeat a whole sentence sometimes, politicians do it all the time in their speeches with what they intend as memorable lines.

Smile whenever appropriate. Be natural: if you need to pause, to look at your notes, blow your nose, or take a drink, go ahead and just do it. Keep track of the response of the audience by monitoring their facial expressions. It helps to make eye contact with two or three persons and to talk to them, as if intimately.

The no-nos? Avoid reading your notes, you should have memorized the gist of your talk. Do not read what shows on the screen, if you are using visuals, the audience is capable of doing it for themselves. Rather than reading, paraphrase what's written, improvise small asides, humorous if possible. Flesh out in an interesting manner the text and images on your PowerPoint transparencies.

And do not forget to end with a recapitulation, underlining your main points in a well-organized and literate manner. Take leave of your audience genially, kindly and courteously. Thank them for the attention they have given you.

REFERENCE: A. Wennerstrom and A. F. Siegel (2003) *Keeping the floor in multiparty conversations: intonation, syntax, and pause.* Discourse Processes 36(2):77–107.

an action sentence 

TITLE

The two main recommendations are to avoid title-drafting by committee; and for the senior author to spend time putting together a long list of tentative titles, at least a dozen. Prune them to a selected few. They will yield a final choice, whether used as is, or modified until satisfactory.

But let us consider a few examples. »Direct observation of electron dynamics in the attosecond domain« is a good title. It does not use too many words. It states clearly the object of the publication. Can we improve it?

»Attosecond domain« may sound redundant. However, its referent is evident, the authors wish to report their observations in the time domain. Were one to shorten the title to »Observation of attosecond electron dynamics«, nothing of substance would be lost. Yet, the modified title is improved. Be concise, it increases readability. Note suppression of the adjective »direct«, which does not add much to the title – except a sprinkling of hype.

Consider now »Massively parallel manipulation of single cells and microparticles using optical images«. This is a problematic title. Its core, manipulation of single cells and microparticles is fine. We understand that tiny objects, in the scale of microns, are being moved. One of the obstructions to easy understanding is the catch-all word »using«. The title would read better with »monitoring« instead of »using«: one understands the action undertaken and reported to have been watched with a powerful instrument of the microscope type.

The other difficulty comes from the technical phrase »massively parallel«. What I understand it to mean is for the miniscule objects to have been displaced, as a group and in the same direction. Instead, why not »Moving single cells and microparticles as large ensembles, monitored with optical images?«

Or, »Optical images for observing large-scale manipulation of single cells and microparticles?« In this last revision, the contrast between »large-scale« and »single cells« provides an interesting trope. It then only suffices to get rid of the redundancy and shorten the sentence to »Optical observation of large-scale manipulation of single cells and microparticles« to create what is, at least in my opinion, a better title.

My third example is »Serum retinol binding protein 4 contributes to insulin resistance in obesity and type 2 diabetes.« It is long. However, to shorten it is not obvious. This is an action **sentence**, always an outstanding choice. Its subject, serum retinol binding protein 4, cannot be shortened, except with recourse to an **acronym**, a move to be banned except in very few cases (such as DNA). The phrase »insulin resistance« is both standard and crystal-clear. Likewise, the two »locations« of insulin resistance, obesity and type 2 diabetes, are accurately and well characterized. The only room for improvement here is with the verb, to contribute. It is a little vague, probably deliberately. One may prefer to be more specific and to rephrase this title slightly: »Serum retinol binding protein 4 bolsters insulin resistance in obesity and type 2 diabetes.«

My last example: »EphrinB2 is the entry receptor for Nipah virus, an emergent deadly paramyxovirus«. The title is somewhat unsatisfactory, it carries two distinct ideas. Why not, then, use two separate sentences: »Nipah, an emergent deadly paramyxovirus: identification of the entry receptor?« »Deadly« is a little surprising, we are used to the adjective »lethal« instead. Choosing between the two is a matter of taste. Nevertheless, those are truly minor quibbles, the original title was very good to start with.

To sum up: clarity first and foremost, concision next.

REFERENCE: M. Alley (1996) *The craft of scientific writing.* 3rd edn, Springer, Berlin, Heidelberg, New York.

it rubs raw 

UNDERSTATEMENT

We have all had the experience. We are in a noisy place, a restaurant maybe. People start raising their voices. A deafening din ensues.

In our time, a resemblance is striking: echoes from the advertising world, whether slogans or images, also try to outdo one another. »X is good for you.« »X is the best in the world.« Moreover, they may intimate at rewards, at personal self-improvement, at a reconciliation with nature, what have you. Their favorite rhetoric is that of the hyperbole – of hype for short.

Hype is not absent from scientific papers. Primary publications resort to it especially. It is found primarily in the **introduction** and **conclusion**. Sentences such as »There has been considerable interest recently in ...« have become near-standard. Other examples include the obsequious »There are few times in a scientist's career that the opportunity arises to describe a major portion of his own work in a journal as important as …«, the one-upmanship of »we start with our discussion at the innermost core of the phenomenon …«, or passing off as a visionary with »the fundamental points in the design of any architecture …«.

The point in such an attitude is to tell the world, »see how great we are«. This is not the way to do it. It rubs raw your colleagues' skins. It meets with their natural skepticism. It may cause them to shrug inwardly. Instead, let the restricted community of your peers praise you (or damn you). Cheap sensationalism belongs in tabloid dailies with their screaming headlines, not in science journals.

Hype tends to creep in naturally under the pen. A good characterization of hype is overindulgence in adjectives. Just like overindulging in sweets leads to obesity, accumulating adjectives bloats a text, makes it bottom-heavy and turns it into failure.

Avoid hype. Go for the understated. »Small is beautiful.« You won't seduce fellow-scientists by shouting your own praise. In the words of a colleague, always, always hype acts to put a minus mark on a person … Be modest. You will be repaid in a growing reputation.

Science is cumulative. Your contributions add to it, one small step at a time – even if it appears to be a major step, from your egocentric vantage point. As Newton himself put it (he was reiterating a metaphor dating back to Bernard de Chartres, ca. 1130), »*it I have seen further it is by standing on ye shoulders of giants*«.

REFERENCE: D. Berntsen and J. M. Kennedy (1996) *Unresolved contradictions specifying attitudes – in metaphor, irony, understatement and tautology.* Poetics 24:13–29.

the backbone of the sentence 

VERBS

A verb is the backbone of the sentence. It needs strength. As a rule, scientific publications almost uniformly include weak verbs. Overuse has weakened the meaning of verbs such as: confirm – control – demonstrate – describe – design – display – emerge – exploit – find – fulfill – implement – involve – consist – form – monitor – observe – obtain – occur – suggest – surround – report – represent – show. Furthermore, again as a rule, such verbs are used in the passive voice, which further drains them of substance.

One of the solutions, anthropomorphic since we stand on two legs, is to hang the **sentence** on two rather than a single verb: »*a major line of investigation ... is to understand the processes underlying the self-organization of matter and to implement them in artificial systems*«. This particular sentence resembles one of those tents held on two masts. Those are the two main verbs, to understantand and to implement. Their difference in meaning activates the sentence, it moves it along from the more passive act »to understand« to the more active »to implement«.

Another crutch is the adverb: »*Figure 2 shows unequivocally that CH_4 re-forming reactions do not change the number of exposed Ir atoms*«. A week verb, to show, calls for an adverb strengthener, unequivocally in this case. But there were other, better options. By turning around the sentence, one rewrites it as, for instance, »CH_4 re-forming reactions do not change the number of exposed Ir atoms, that is the lesson from Figure 2«. Editing improves it further to »CH_4 re-forming reactions do not change the number of exposed Ir atoms, as Figure 2 tells us«.

The best solution by far, rather than recourse to two verbs or to adverbial-strengthening, is a strong verb. How does one recognize such a verb? By its infrequent use in scientific publications. How does one find it? By looking it up: the number one repository for vocabulary, in the English language, is *Roget's Thesaurus*. Get yourself a copy of this inexpensive little book. While a few of its idioms, nouns, adjectives, verbs, are hopelessly obsolete, it will nevertheless provide you with an enrichment of the lexicon.

For instance, paragraph 525 of the *Thesaurus* offers quite a number of alternatives to »show«: to manifest, express, indicate, point out, bring forth, bring forward, set forth, expose, produce, bring into view, set before one, hold up to view, lay open, lay bare, expose to view, set before one's eyes, show up, bring to light, display, demonstrate, unroll, unveil, unmask, disclose.

REFERENCE: R. Graves and A. Hodge (1990) *The use and abuse of the english language*. Paragon House, New York

OTHER REFERENCES: the first quoted sentence is from A. Petitjean et al (2004) *Angew chem int edn engl* 43:3695–3699; the second from J. Wei and E. Iglesia (2004) *Angew chem int edn engl* 43:3685–3688.

to paraphrase creatively 

VISUALS (FOR A LECTURE)

To seduce through the ears and to appeal to the eyes, those are your tools for a talk using visual aids. What are those? Typically slides or transparencies, they alloy images and text. Sometimes, they consist only of images, or just text. Aim at about a slide per minute, about 60 for a one-hour talk; for transparencies, every two or three minutes, which translates to about 25 for a one-hour lecture.

Include in your presentation the recurring indication of the advancement of your talk. It is important that your listeners be given such information: how many more parts to the talk, how far from the summing up, when is it going to be over? Show a table of contents at the start, return periodically to it to indicate what remains to be covered.

My first rule is the most difficult to implement: make your visuals and your speech complementary. Their occasional redundancy is good. But avoid duplication becoming systematic. For instance, avoid reading aloud what is written on your slides. Your audience can do it much more quickly silently. Paraphrase the words on the screen. This is easy. However, to paraphrase creatively, in a witty and seemingly spontaneous way, is quite another ballgame. How to achieve it? By practice.

The second rule is substraction. Once you have assembled a first draft of a presentation – using PowerPoint for instance – delete at least 20% of the material from individual slides/transparencies. You will find them quite a bit more effective.

For the text, select a single font for readability. However sexy they may look, sanserif fonts such as Arial or Geneva are hard to read. The serifs are the small slippers which letters bear on their feet. Their implied horizontal continuity helps the scanning eye. Personally, I favor a serif font such as Palatino or Times.

A fourth rule is to strive for continuity, built-in unobtrusively. Be attentive to the dominant color in your sequence of visuals. For instance, you might start with a light pink background and, rather than sticking to a uniform background for the entire presentation (often advisable) gradually work your way to a deep red. Speaking of such color cues, be attentive also to their psychological, i.e., cultural connotations.

My fifth rule is to maximize the iconic versus the textual content. Tables of numbers are an absolute no-no: only display your data with plots. Remember to show just a fraction of your data (Rule Two).

And what should your images be? Obviously, they need visual interest. Show images, which are a compromise between harmony and a jolt; the aesthetically pleasing and the informative; the obvious and the enigmatic. The mind enjoys having to do a little work – a little, not too much – when reading the picture. This is good. To gain understanding actively rather than as a passive viewer gives pleasure.

Include some whimsey. Don't hesitate to be a little outrageous. Back in the 1960s, before permissivity became widespread, during a conference of advertisers one of the speakers introduced a few mineralogical pictures among his slides. They just flashed through, with no comments on his part. His talk was a major success. In other words, keep your audience alert by, occasionally, showing it the unexpected. As in the above anecdote, the occasional surprise material, shown without comment, will spice up your talk.

Not only fantasy, beauty is also an essential ingredient. Ideally, each talk ought to include one piece of art capable of being exhibited separately, because it can stand on its own.

What about the twin projector mode, setting-up a counterpoint between two images projected together, next to one another and occasionally overlapping? Such an ambitious undertaking is worth the additional effort – and money, if you have the show prepared by professionals, which I recommend – only if the actual product shown to your audience is perfect. Moreover, it makes for a more passive audience: are you sure you really want it?

REFERENCE: E. R. Tufte (1997) *Envisioning information.* Graphics Press, Cheshire, CT.

gloomy and disquieting 

VOCABULARY

The atmosphere within your piece is determined by your wording. If you want it to be gloomy and disquieting to your readers, then just go ahead and make systematic use of complicated words. Conversely, if you would rather come up with an airy piece, one which readers will traverse self-confidently, then make sure to use short and simple words. Brevity is golden. When in doubt, choose the shorter word. Use words that come naturally to you, words you would use in speech.

The English language has a huge vocabulary to draw from. A reason is that, following the invasion by the Normans in 1066, it became the juxtaposition of two languages. Latinate forms, imported at that time with the French language, coexist in English with Saxon words. The tradition has endured. The former, multisyllabic words abound in scholarly works; whereas the simpler, often monosyllabic Saxon words tend to be used in daily life.

A good rule is to comb your text for complicated Latinate words and replace them with their Saxon counterparts. This will lighten your prose. Examples? fast/rapid; speed/velocity; cloudy/nebulous; hint/suggestion; handbook/manual; settlers/colonists; get/receive; put together/assemble.

A second rule is sparingly use of recent **neologisms** in your discipline, if at all. Yes, I am telling you to renounce being trendy. You want your work to become a classic, right? Then, write in the classical manner. Eschew complicated words. Some of our colleagues have introduced them for their own reasons, in which one-upmanship has had a part.

Which is not to say that you ought to avoid totally using learned, seldom-used words. Any reader loves to learn a new word. Why? Young children, it is said, learn ten new words a day. In adults, the learning process slows down. But it does not come to a complete stop. Many of us keep learning, including new words. And this is enjoyable. A writer such as Stephen Jay Gould was very much aware of this urge on the part of readers. He provided them with the occasional mouthful word, one which he was astute enough either to define, when he introduced it, or one whose meaning people could readily guess simply from the context. My rule here is to put in – »put in«, rather than »insert« – such a nugget every 2,000 words or so.

REFERENCES: the switch from Latinate to Saxon is associated with Basic English, which C. K. Ogden introduced, rather successfully, in 1923 in his book with I. A. Richards (1965) *The meaning of meaning.* A good place to look-up the right word is *A dictionary of modern english usage.* 2nd edn, H. W. Fowler, E. Gower, reviser, Oxford University Press, New York.

a good designer

WEBSITE

This is the contemporary calling card. Henceforth, you will be identified more and more by this presentation. Note, however, the role reversal. Yesteryear, you would present your calling card upon meeting a person. Now, you are not calling on anyone with your website. The whole world unbeknownst to you calls on you. In addition to the calling card, the website fulfills a number of other functions. For instance, it supersedes the list of publications which, of course, it partly includes.

How should you adapt to this switch in self-identification? What should and should not be posted on your site? Needed ingredients are a **capsule biography** of yourself, a selected **bibliography** of your publications, **abstracts** of selected papers and lectures, a few publications in full, chosen for their high significance or quality, a brief summary of the topics you are currently working on, the composition of the personnel in your laboratory, a guestbook (but only if you find the time to read it on a weekly basis), also perhaps a counter for the number of visitors.

The requirements are severe, which turns the creation of a website into a large investment in time and money. The main requirement is visual attractiveness. Take examples from magazines. Like them on the stand at a newsagent, you are in competition with many others. An unattractive website won't be visited. This is surely not your intent in putting one up for view and consultation. It has to be illustrated, for which you may well need a professional iconographer – at least a professional touch.

No wonder if graphic artists, made redundant by software available to everyone (such as Photoshop), did find another niche for themselves and are now designing websites. Yet, uploading a website calls also for other, complementary talents. It needs people capable of programming into various computer languages. It needs an expert in search engine optimization, for your site to be located easily, close to the top of the Google page – after your name has been typed in. It needs a person for key word research. Another specialist is required for usability studies. One of the mandatory subdivisions is links, they need to be built in. The site starts with a clear and simple index cover page, which is the responsibility of the graphic designer.

Once you have hired an outfit to design your site, be prepared for being called upon to provide information, repeatedly. The back-and-forth in in the designing and building process is analogous to that with an architect and a contractor building a house.

These specialists will start by giving you a wireframe, i.e., a skeleton website indicating all the navigation, function and contents elements to be included in the final product. The wireframe lacks any graphic design elements. It serves as a blueprint for the later construction steps. Using the wireframe together with the sitemap, you and the designer will plan the textual contents. Writing them will be your biggest workload, it is time-consuming. A good designer will be attentive to your expressed needs and desires.

What is also time-consuming is the putting together of a draft version, then submitted to you. It will be improved with your feedback. And you will go through another cycle of submission, corrections and revisions and improvements.

After your site is uploaded, it needs to go through a final, testing phase by potential users, prior to being made fully functional.

REFERENCE: P. v. Dijck (2003) *Information architecture for designers.* RotoVision, Hove, East Sussex U K.

information overload 

WIT

Is not science utterly serious? Why then does it need a witty form of communication?

Because of the information overload we are facing. Our audiences privilege attractive messages. This holds true not only for the general public, but for our peers as well. Not only is a wittily presented piece lighter and thus easier to assimilate, it is also more likely to be remembered and recalled. Your aim is to combine being informative and entertaining.

How best can this be done? Can it be learned? My advice is to model yourself on speakers and writers whom you admire for their wit. They bring a playful attitude to their task. While respectful of their public, they seem to be fully at ease. Their personality shines through. They make playful use of words. Their very vocabulary is fun. It is fun because they spring little surprises. An occasional word in speech or from their pen is unexpected. It may borrow from everyday language, bringing fresh air into what would qualify otherwise only as specialized and highly technical scientific terminology. The unexpectedness may stem conversely from it being a word used relatively rarely even though readily understood.

But there are many other instances of wit apart from words. Wit is a state of mind, a mood. It has to be effortless, it has to sound natural. Cracking a couple of jokes at the start of a talk may be standard behavior for a salesperson or a politician, it is beneath the dignity of a scientist, as formulaic and definitely an unseemly kind of behavior.

The art of being witty thus entails being spontaneously bubbly, without letting it crystallize into distracting witticisms or puns. A good start is to put yourself into the shoes of the storyteller. You are presenting a narrative. Do so in a good-natured, seemingly spontaneous way. Practice doing so up to the point when you feel yourself in danger of losing it, of making yourself giggle from what you have just uttered. This is a litmus test of your wit.

A message which is not also entertaining is like one in a bottle dropped into the ocean. Only by the greatest of luck will it reach its recipients. You want to catch attention by being amusing, being careful neither to clown nor to mask your message with too pretty and too funny an envelope.

Get a handle on this talent by interspersing in your talk – this is done more easily in spoken than in written presentations – the occasional anecdote. Such inserts, almost mechanically, will infiltrate the rest of your talk with their light-hearted and genuine spirit. In so doing, you will grasp a basic notion, no effective communication without a modicum of amusement and entertainment.

REFERENCE: W. Strunk, Jr. and E. B. White (2000) *The elements of style.* 4th edn, Allyn & Bacon/Longman, New York.

rise to the challenge 

PART I

ADDRESSING PEERS

GENRES

AFTER-DINNER SPEECH

A tough assignment indeed of the kind to be wished on your worst enemy. Why is it difficult to address a group of well-fed people who ought to feel gratified and hence benevolent? Because your slightly enebriated audience has a single wish, to be entertained, plus a single expectation, to be bored into sleep. You are warned, these people very easily will become drowsy, or leave.

Therefore, such an assignment has to be prepared with great care. Your standard talk won't wash. For one thing, you can't lean on your usual crutch, the audiovisual, plus the room is not really suited for that purpose. It is designed to serve a meal to a large number of guests. Even if there is projection equipment, deny it from yourself, you will be better off with **body language**: now that you are a personality on the dais, and digestion is turning them into young kids ready for their nap, your listeners want to get to know you as a person. Making eye-to-eye contact with each of the diners will be quite a bit more difficult than in a standard lecture hall. Nevertheless, rise to the challenge. Be personal if you can. Memorize a few names from the badges or the place settings and refer by name to some of the individuals present.

Laughter is your greatest asset. It is also your most likely downfall. It is awaiting the slightest signal on your part, deliberate or inadvertent. You will have to include a few jokes, especially at the start. They are near-compulsory. Which of course does not preclude a serious topic. Just as in a play of Shakespeare's, the comic elements coexist with the serious. You want to carefully avoid histrionics. Don't try to pass as a clown or a buffoon. If you do, whether deliberately or not, none of what you say will register.

You are competing, especially at the beginning, with private conversations. They started before you took the floor, and well-lubricated with alcohol, they will take precedence.

How to counter? Bring up, at the start, some object or contraption. From curiosity, the audience will focus on you. Vary your speech delivery, in loudness and in pace. Take the audience on a bit of a roller-coaster. It is important they should feel led, held attentive by your narrative and your delivery, better yet, fascinated by you as a public speaker.

The speech itself must exude simplicity. In its organization. In its ideas. Make few points and make them forcefully. Recapitulation at the end is even more essential than usual. Sprinkle your talk with the salt of wit, with the spice of comedy. Interspersed during the speech, they will hold your audience captive.

You need to be perceived as a dynamic personality and a gifted communicator. Avoid being static, keep moving both physically and figuratively, give your audience a strong push forward.

Have you ever watched a downhill skier in a race? You need to, likewise, appear to be in a rush and yet totally at ease.

REFERENCE: P. H. Andrews (1985) *Basic public speaking.* Harper and Row, New York.

the implicit anticipation 

BOOK CHAPTER

To be called upon to contribute to a collective monograph is a tribute to one's expertise. At the same time, the invitation to write a chapter contains the implicit anticipation that you will bring to bear qualities of fair and neutral outlook, in addition to fine writing.

Follow the instructions received from the editor. Your very first step is to gather the necessary documentation. It requires that you contact colleagues and libraries for photocopies of material not yet in digital form on the Web. This is also the stage to select material needed to quote or reproduce in your chapter, typically **illustrations**, graphs or diagrams. Mail out the requests for permission, which the publisher will need for copyright purposes, at the time your chapter has been edited and is ready to be sent to the printers.

The **bibliography** in many ways will be the most important part of your chapter. Make sure that it is accurate as well as comprehensive. In preparing it, do not hesitate overextending your reach. Gather about 120% of the references you will finally cite. Some will fall by the wayside during the writing stage when you come to the decision to narrow your scope somewhat with respect to what you had initially envisioned.

Do not overreach the assigned format. Once the first draft has been prepared – not on the eve of the deadline for submission, one solid month before – prune it ruthlessly. Your chapter will be so much the better for it.

Should you have questions, consult the editor. In any case, offer an unsolicited progress report to the editor during the writing of the chapter. If he or she does not hear from you, it is a sure sign that you are a delinquent author and that you have not yet started your task. Your reputation may suffer accordingly.

Now, a good rule of thumb. I offer it as a truly discerning criterion regarding material to include and what to leave out. Your chapter should contain everything that an academic might use to prepare a class for graduate students. Exceeding it will only needlessly fill up pages. Have some consideration for pinetrees turned into paper pulp to record your words for posterity: Are you absolutely certain that this particular paragraph is truly needed?

Do not take the opportunity to advertise your own work. Take the view from Sirius, and put it in context. Be fair to your colleagues and competitors. Attempt a bird's eye view.

Yet, you also need to show mastery over the whole area that your chapter covers. Do so subtly. Point out nonobvious connections between different parts of the field. Indicate observations or experiments begging to be done which the field still lacks. Commend quality and original work from other laboratories.

Your chapter has to be authoritative. In addition, it must carry your imprint. Include in your chapter at least one element which will make it personal and idiosyncratic, let your personality shine through. If you have not yet found your own voice, this is a good time to start.

In one or two pages at the end spell out the complementary work which in your opinion needs to be done. But don't try to forecast, your predictions might come back to haunt you.

REFERENCE: J. R. Trimble (1975) *Writing with style: conversations on the art of writing.* Prentice-Hall, Englewood Cliffs NJ.

proselytizing in print 

BOOK PROPOSAL

The trap for the unwary is the great ease with which a technical book is publishable. »Publishable« should not be construed as »easy to write«. To publish is easy because science publishers have taken university and public libraries as their hostages. Almost whatever the topic, however astronomical the cost to an ordinary reader's pocket, they will sell between 500 and 1,000 books worldwide. They build their budget on this comfortable cushion and make hefty profits. You will be performing a public service by opting for another publisher than the two or three giant multinationals which have a near-stranglehold on this highly peculiar »market«.

Another recommendation, before you start drafting a proposal, is for you to visit a university library sporting open stacks. Look over its shelves. Find out for yourself how quickly most technical books become outdated, superseded and obsolete, dead wood literally speaking.

Go ahead if the view of such a cemetery is no deterrent. Such a decision from you shows you do not lack self-confidence. Your better judgment does not dampen your enthusiasm for proselytizing in print. In brief, you yearn to write and get out a book.

The best time for doing so is when you have become a leader in a subdiscipline, and before you switch to another field. Rewards to be expected? As a rule, neither money nor prestige within your restricted community. Your peers will be envious, if you are successful. They won't give you credit for stepping out of line. They keep to primary publications. Thus, they will resent your nonconformism.

A big benefit to you is that, after you have had a few (three to six) technical monographs under your belt, it will greatly help you to obtain publication in other genres – such as popularization for the general public.

The hidden agenda for writing a proposal, beneath the obvious purpose of winning a contract, is to clarify the project in your own mind. You will, then, present it concisely, on less than a page, showing how beautiful your mastery of the topic is, and how novel your book will be, in approach and in outlook. As I have intimated, this goes above and beyond the needs of the publisher. But you owe it both to yourself and to the longevity of the book.

Several annexes are needed. Make a study and write down the bibliographic information for books, already on the market, in the same general area as the proposed. Such a segment ought to be in the style of a little bibliographic essay: for each **title**, indicate its strengths and its shortcomings. If there are colleagues who, to your knowledge, are engaged in a similar venture as yours, provide also such information.

Indicate the names and addresses including their e-addresses of three to five experts in the field, who might be called upon to review your project. Insert an article or two you have written on the same topic, which you plan to cover in more depth and breadth in the book. Insert also, as part of your proposal, excerpts from reviews of your earlier publications. The proposal will be complete with inclusion of a short biography and of a list of your publications – in abbreviated format, citing explicitly only those relevant to the proposed book. Do not fail to indicate how you can be reached: in writing, by telephone or by **e-mail**.

When your proposal yields you a contract, make a careful note of what an editor – preferably not the acquisitions editor, but the person who will be working on your manuscript – will have to say about your project. This advice will be precious.

REFERENCE: C. Turk and J. Kirkman (1977) *Effective writing: improving scientific, technical and business communication.* 2nd edn, E. & F. N. Spon, London UK.

and perhaps memorable 

CONFERENCE PRESENTATION

Scientists gather to trade information about their latest results. At least this is the ostensible role of a conference. There are other purposes, no less important, such as determining who belongs in the restricted community of peers, gossip, information about grants and jobs, meeting old friends from graduate school, and so on.

Assume here that your paper was selected for oral presentation on the basis of a submitted abstract. Carefully choose the format of your visuals. Currently (2005), the PowerPoint software is ubiquitous. Be aware of its shortcomings. Edward Tufte, a Yale professor who is the leading authority, worldwide, on graphics, has issued a booklet with a scathing criticism of PowerPoint.

In case you need to bring up the same detailed material repeatedly – this may be a graph, a table of numbers, a quotation – it belongs in a handout to be passed around as you begin.

Your visuals should combine information of two quite different types. Show the main points in your argument as text, and use images pleasant to look at. To select the former, make a list of keywords in your written paper. To select the latter, go for the colorful, the visually appealing. Stick to the simplest means. You can do worse than using a cartoon pertinent to the topic as a model. Think of your individual images as billboards, and do not cram them with information. Stick to the essentials.

Your **visuals** contain text. Even though you have boiled it down to the nitty-gritty, do not treat it as if it were the gospel. Above all, do not read aloud what is written on your slides. Your audience knows how to read. Your spoken remarks should complement the words displayed. By the same token, do not read your paper unless your main goal is putting your audience to sleep. Memorize your opening and closing **sentences**, so that they be articulate, literate and perhaps memorable. In-between, present off-the-cuff remarks which will sound spontaneous – all the more so that you will have carefully rehearsed your presentation a few times beforehand.

Whatever the time allotted to you, cut it short by a couple of minutes. The audience's gratitude will be palpable. It will ensure that they will be on your side when the discussion opens.

Adjust to the program, so that you fit in. This sounds obvious and yet 80% of presentations do not heed this requirement. Besides, have enough flexibility to adjust to last-minute changes. For instance, if worse comes to worse, be prepared to make a blackboard-and-chalk presentation. Listen carefully to the speaker ahead of you who offers a foil for your own talk. Pitch your voice and your delivery for pleasant contrast.

You are offering both your results and yourself to criticism. Be neither defensive nor aggressive. Remain modest. Be attentive to what you are being told. Take the time to jot it down to aid your answers and as a reference to the comments made. Give genuine respect to the authorities, the chairperson, and listeners who may collectively know considerably more than you do.

Heed the Golden Rule. Start your talk by telling people what you are about to say. Say it. Conclude by telling people what they have just heard, i.e., summarize your main points in no more than 25 words.

REFERENCE: A.M. Penrose and S.B. Katz (2003) *Writing in the Sciences: exploring conventions of scientific discourse*, 2nd edn, Longman, Harlow, Essex UK.

an astute oblique way 

EDITING PROOFS

This task requires that you be at your best behavior as a good citizen. A spirit of cooperation demands introducing as few changes as possible into the text.

This requirement runs contrary to what may be your spontaneous instinct, that is, to restructure, remodel and rewrite your contribution. You may find some of your writing to have been unfelicitous, clumsy, wooden or difficult to read. The impulse will then be to redo it. Consider however, such an unhappy experience as a learning process. Teach yourself to write cogently prior to submission. Afterwards, it is too late.

There is a single exception. it concerns new material obtained after submitting your paper. If you feel that readers ought to be provided with the new information, then it should consist of a single additional paragraph to be entitled »Note added in proof«, which you will tack to the corrected proofs at the same time that you supply a copy to the editor (by e-mail, presumably), out of elementary courtesy. Introducing extraneous material at this very late and postmature stage, let me emphasize, is only acceptable if it is of true service, not to yourself, but to the scientific community. If you opt indeed for that route, the extra text ought to conform to the style of an abstract: brief and to the point.

There is an additional point. What if a referee has insisted on a change that you strongly object to? It might be the rephrasing of your title or subtitle. It might be introduction of a bibliographic reference which you feel is totally irrelevant.

At this stage, the temptation is to override the referee. Think very carefully before doing that. Were you to decide to ignore the authority of a reviewer, find an astute oblique way of doing so without setting up a confrontation.

Which brings up a more general point, to help ensure that the editor is agreeable to the whole set of corrections entered in the proofs. Write this individual of the actions you have taken, excepting none and offering justification for each. Such a recapitulation is a sign of respect. You can gain editorial support only by being totally transparent and informative with respect to your actions. Being sneaky, doing things in a covert way will only get you into trouble.

Make all your corrections and/or additions in the most legible manner. For modifications of more than a few words, append a printout of those sentences as you would like them to appear. Use the prescribed printers' symbols for your corrections. It is a good idea to enter them first in pencil and to finalize them in ink at the very end, just before mailing them out.

Make a photocopy of the before and after for your files.

REFERENCE: Editorial Staff, University of Chicago Press (2003) *The Chicago manual of style: the essential guide for writers, editors, and publishers.* 15th edn, University of Chicago Press, Chicago, London.

a conversational nature 

E-MAIL

This mode of communication has now been around long enough to have degraded quite a bit; commercial purveyors of spam are not the only culprits. I shall try to provide here a few practical guidelines, to put in also a word about e-mail etiquette and about some technical constraints.

As a means for instant communication, irrespective of location and time zone, e-mail is near-ideal for the worldwide scientific community. For one thing, scientists were at the origin of both the Internet and of e-mailing, with the ARPANET in the US, and at CERN in Geneva. Thus, it should be no surprise that it has continued to be the standard mode of communication among scientists, for more than 20 years now. I have written myself several articles in collaboration with colleagues across the world, without needing a single face-to-face meeting: e-mail allowed this small feat.

E-mail may be compared (not in cost) to the telegram, which went out of fashion in developed countries several decades ago. Like the telegram, it has a position in-between a formal letter, otherwise known as snail mail, and a phone call. Like written correspondence, it consists of a text; which in principle can be archived, as the hard copy or on a variety of supports. Like the phone call, it is informal and often of a conversational nature. However, feelings and emotions are considerably more difficult, more delicate to convey, than with a phone call. Conversely, e-mail is vastly superior to the telephone in the density of information transmitted per minute, by one or two orders of magnitude.

The first consideration is the need to adjust the impedances, so to say, between the sender and the receiver. This calls for proper and explicit reference to the previous message(s) in your exchange. Do not assume that the recipient, who may be besieged with several dozen e-mails daily, will know instantly what you are writing about. Make it very clear.

You have no idea, as the sender, of the frame of mind, mood and intentions of the receiver. Accordingly, you need to be very careful about contextualizing your message carefully and courteously. Likewise, compose subject line with imagination. It needs to be accurate, to the point, explicit and terse. Watch out: you do not want it to be confused with spam, and the attendant, perhaps even automatic, deletion of your message.

If you address a message to multiple recipients – a procedure in principle to be avoided, which needlessly increases e-traffic and unsolicited verbiage across hyperspace – their e-addresses need go into the cci, not the cc box: you are not supposed to divulge e-addresses you are privy to. While on the topic of e-mail etiquette, do not enclose large files (such as photographs) without first asking for permission to do so. You risk antagonizing a colleague by cluttering his or her electronic mailbox.

Coming back to the analogy with a telegram, keep your message brief: 25 lines of text, as a rule, are a maximum. Stick to short **paragraphs**. Make liberal use of blank space in-between paragraphs – this will give your text some breathing room. Keep these lines under 70 characters-long – the technical reason is that otherwise your recipient risks getting some nonsense symbols. Back in the early nineteen-eighties, when e-mail was conceived and formatted, only 128 characters were allowed for transmission between computers. This restriction still obtains nowadays. Hence, do not be surprised if bold type, italics and other special symbols in your message do not get through to the receiver. E-mail is not an excuse not to watch your grammar, punctuation and spelling.

To sum up: with e-mail, we are back to what Æsop wrote about the tongue and language, during Roman antiquity. It can be the best of things. It can also be the very worst.

REFERENCE: D. Blum and M. Knudson (2005) *A field guide for science writers.* Oxford University Press, Oxford.

into a dream version 

EDITING

Your profession is scientist, not editor. Nevertheless, at some point or another your career may well involve an editing task. Science is inseparable from publication, hence your likely foray into editing.

What is editing, what does an editor do? In a cursory answer, the editor is midwife to the writer. The editor assists in the writing process, makes sure that a piece is readable and conforms to the standards of idiomatic English syntax, **vocabulary**, spelling and punctuation. An editor sees to it that the scientific terminology is apt and accurate.

When a graduate student writes a Ph. D. dissertation, it is the duty of the supervisor to become the first reader and help turn the first drafts into an acceptable final product. There are many other instances calling for an editing gig, whether on the academic or on the industrial side of a scientist's career.

At this stage, I'll stick my neck out and offer another definition of editing. The editor of any text is a reader who strives to turn any submitted text into a dream version. Expressed yet in another way, editing any text aims at streamlining it, making it follow all of the rules of correct expression, without altering the tone of an individual voice. Does that sound easy? It is not that easy. It does become easier and quite natural with some experience. It is not that difficult either. As one reads a text, the needed corrections jump at you. It is very time-consuming though. You may spend an hour editing just two or three pages.

I have dealt so far with amateur editors such as ourselves. There are also professional editors out there and we often deal with them. These are people in the publishing business whose job it is to work on incoming manuscripts and turning them into books. They come with a variety of hats: the acquisition editor, responsible for bringing-in promising authors and their manuscripts; the developmental editor, who helps turn an inchoate idea or a poorly organized manuscript into one which will interest an agent or a publisher; the editorial director, who steers the whole output of a company and who also sees that the publication program is balanced and consistent with the existing list; the copy editor, finally, who prepares the text for the printers and who checks the accuracy of each statement, quote or reference. A professional editor is someone with such a command of the language that s/he will turn a text into a readable and correct piece, a wizard with vocabulary and syntax. A professional editor keeps in mind two versions of a text simultaneously, the actual piece in front of the eyes and the metamorphosed end product, prefigured in a manuscript. It is an uncanny talent to have such a vision and to follow it through, carrying out what needs being done to turn the Cinderella-manuscript into a princess.

In-between amateur and professional editors, I should not omit mention of scientists with extensive experience in editing from being associate editors or the editor of a journal, from having edited a number of scholarly monographs, … Those have borne the responsibility for a publication and its quality, in whole or in part.

To sum-up: we owe a lot to editors who have made our prose acceptable for publication; if called upon to help others in like manner, we may want to turn our indebtedness into service.

REFERENCE: M. Alley (2000) *The craft of editing: a guide for managers, scientists and engineers.* Springer, New York.

these laggards 

EDITING A BOOK

It is likely to be a collective volume. You felt there was a need for a book on this particular topic. Now you have secured a contract from a publisher. With the legal document or as a part of it, are included the tentative title, the length of the text (publishers have no flexibility on that issue), the list of contributors and of their chapters. This initial part of the venture is easy.

Make sure to put together an interesting table of contents. Most scholarly monographs are monotonous. They suffer literally from being monotonal: their material is organized from a single principle, linearly. Stereoscopic vision brings in the depth which monoscopic vision does not provide. Use two approaches, orthogonal to one another, such as history and geography; contemporary vs. historical; local vs. global; experimental and theoretical; etc.

In my experience, having edited a number of monographs crisscrossing of a field, of any field with such a grid works well. Besides a pair of complementary approaches, other qualities of your proposed book ought to include, in order of decreasing importance: 1. originality, which may follow from bringing on board an odd man or two, with radically different perspectives; 2. overall harmony and consistency, which will require you to fill the voids and probably write yourself the missing sections; 3. comprehensiveness. The latter feature is not that important. An innovative presentation of a field is more stimulating, more productive than any encyclopedic treatise.

Now that you have put together your program, fulfilling it is easier said than done. A rule of thumb is, for any dozen authors having signed the publishing agreement, you will be beset with two or three delinquents. These laggards will fail to turn in their assignment on time. It is wise to brace yourself for that certainty by commissioning more authors than needed on paper, say 15 for a twelve-chapter book.

Be tough. My rule is to keep pestering the authors. They know full well, you do too, they won't start writing the first word of their contribution until the due date. They will come up with rich, imaginative excuses. However, you need no excuses, you need a completed manuscript.

Insist that each author supplies you with an outline of the contribution upon signing the publishing agreement. This document, optimally two- or three-pages-long, should list the subheadings for a chapter and should be explicit in the tack to be taken. If not forthcoming, this is a sure sign of more trouble ahead. It may be a good idea to drop this particular author at that point. Alternatively, crack the whip. Ask for the outline to be e-mailed to you the next day or else …

You need to be at your authoritarian worst. Your stable of authors will behave like children. You will need to send them regular reminders, on a monthly basis if the book is a year in the making, to the effect of »please sit down and do your homework«. In my experience, a good inducement for them to do so is to circulate an announcement, say about 40% from the stated deadline, to the effect that you have already umpteen chapters in hand, some of which are already undergoing their post-editing revision. Whether the statement is accurate or not is immaterial. The ploy works.

Once the chapters are in, **editing** them, if a time-consuming chore, is not hard. The authors will approve your interventions. They will go along with your suggested changes all the more readily that they are eager to see their prose in print and hold in hand the published book.

You will provide a **Preface** which, basically, is an expanded table of contents. Commend each of the authors for some positive feature. Offer your personal outlook on the subject matter, its past, present and future.

Last but not least, the **Index**: do it yourself. Again, this takes considerable time. Unless you do it yourself, it will be close to useless to the readers.

REFERENCE: D. Worsley and B. Mayer (1989) *The art of science writing*. Teachers and Writers Cooperative, New York NY.

a network of informants 

EDITING A MAGAZINE

To be called upon to edit a magazine for the readership of one's fellow-scientists, whether within a restricted community (*Ambix*, for historians of chemistry), a discipline (*Chemical and Engineering News*) or scientists in general (*Nature* or *Science*), is a great honor. It is a full-time job. If you accept the task, you will need to take a long-term sabbatical from your work. Getting back to it eventually, after a lapse of several years, may be tricky. Think carefully before accepting.

The main qualities you need to develop and rely upon are imagination and sound judgment about people. Your most important task is to have a vision. Your objectives are what you want this particular magazine to turn into under your stewardship. You need an intuitive perception. Express it to yourself. Make it coherent. Set it down in writing. Moreover and more difficult, hold on to it and follow it tenaciously.

Your second most important task is recruiting your team, an art director and associate editors. Try to get the people who hired you to give you a free hand in doing so. To work with an already existing staff after replacing an outgoing editor is trickier. This calls for treasures of psychology and diplomacy.

Be yourself. If you are incapable of being original, this is not the job for you. Do not follow the trends. Make them.

Work at translating your structural notions into visual ideas. Your art director is your most valuable partner in helping you do so.

Planning the issues is a collective effort. Make sure to inform your whole staff of what you have in mind and want to accomplish. Set-up regular staff meetings to discuss the state of preparedness of each issue in the works. Encourage initiatives from your coworkers. It may be worth your while to renounce an idea of yours even if it is superior, in favor of one originating from someone on your staff. The boost in motivation will amply justify it.

You will need a new dummy every three to four years. This is the full prerogative of your art director. You walk on a tightrope by entrusting this person with the responsibility, while encouraging everyone else on the staff to chip in with ideas. Indeed, the end result has to avoid a patchwork, the planning by committee, it has to be the creation from a single individual. At the same time, each and every member of the staff ought to feel personally involved and proud of the new format.

To a large extent, you will likewise have full responsibility for the subject matter in each issue. How to choose it? How to plan ahead?

For these purposes, you need to quickly build a network of informants among your fellow-scientists: people whose judgment you trust, with their finger on the pulse of science in their specialty, well-informed on what goes on in other laboratories besides their own. They will help you identify prospective authors. A good rule is to go foremost for the quality of science, when unable to grab an outstanding scientist who can write. The necessary rewriting will be up to your staff.

REFERENCE: P. P. Jacobi (1997) *The magazine article: how to think it, plan it, write it* Indiana University Press, Bloomington IN.

acknowledge merit 

EDITING FOR A JOURNAL

Have an exalted view of your responsibilities. Do not restrict your allegiance to the journal only. It extends to the scientific community as a whole. Yet more important, it involves your conscience. You were entrusted with this position because of your experience and expertise. To be of service to the community is what it entails – not a self-serving function in which case you are headed for disaster. Do not expect any rewards except an increased reputation, an increased visibility and respect from your peers, and the significant if immeasurable one of increased worth as a scientist. If you are a good letter-writer, you will appreciate expanding the circle of your scientific friends through correspondence with authors and referees. It has a steep learning curve. Let me give an example of the benefits: from honing the writing skills of your fellow scientists you will, slowly but surely, improve your own writing ability.

What your fellow-editors do is their own business. Be independent and autonomous in your judgments and decisions. You are upholding standards of excellence. Do not compromise them.

The bureaucratic part of your job consists of a set of routines, such as mailing out a letter to acknowledge receipt of a just submitted manuscript. The original part of your job draws on your familiarity with the field. Arguably, your most important task is choosing referees (also known as assessors). That they be competent goes without saying. Other criteria include their past record as referees: were they prompt in response? Did they discharge their duties in a responsible manner? Did they supply a well-argued and detailed report? Was it critical in the best sense of the word, combining the positive and the negative?

You may also want to give yourself the luxury of a nonobvious choice as a referee. This might be a past leader of the field, having moved on to other interests. This person can provide you with a neutral, objective and authoritative judgment.

In the same spirit, striving to give yourself the best advice, do not hesitate to consult other colleagues, whenever you receive a perfunctory referee report, whether positive or negative. It belongs in the waste basket, and if this is not a first occurrence you will need to scratch the name of this particular person from your list of potential referees.

When the referee reports have come in, you will have to make your decision as to accept or turn down the paper. Do not send a standard rejection letter. It has to be personalized. Acknowledge merit, while leaving not even a glimmer of hope to the authors.

If conversely you signify acceptance conditional upon changes recommended by the referees, invest time in the scrutiny of the revised manuscript when it is returned to your office. Some authors will try to get away with cosmetic rather than substantial changes, and your full alertness will be needed.

This is the stage at which you will need to give the paper its final check for the quality of the English and, most probably, engage in copy editing the manuscript into a final state suitable for sending it on to the printers and for posting on the Web.

REFERENCE: M. Alley (2000) *The craft of editing: a guide for managers, scientists and engineers.* Springer, New York.

libraries are vital 

EDITING A JOURNAL

You carry the responsibility of disseminating new knowledge and ideas. Thus, you have several constituencies. Authors and referees, who assess the quality of the work submitted for publication are drawn from the ranks of the discipline – and are to some extent interchangeable. Readers also belong to the same group, perhaps slanted more towards younger scientists who use articles in a journal as an educational tool to help them become proficient in the craft. Libraries are vital. Not only will they serve as your ultimate repository; they are also the gatekeeper. Their willingness and their ability to pay hefty institutional subscriptions often decides the commercial viability of a journal. Publishers, whether commercial or a non-profit professional society, are business-oriented. Their dominant interest is good and financially-sound management. Some journals, very few, do not have any such financial worries.

You are responsible for fine-tuning the dosage of facts and of their interpretation. A mere accumulation of facts, which feed into data banks, is not yet digested and needs interpreting. This is best done by minds experienced at weighing the facts. These are people widely informed about their profession and judicious in their comments. You are one of them. Use of the adjective »critical« is compelling here. An editor must bring to the job a critical mind, i.e., in the etymological sense of the word, the ability to sieve through the material discriminately, retaining what is new and valuable, leaving out the sub-par or merely imitative work.

Not only discriminating in judgment, a good editor must also be liberal in acceptance of a diversity of styles. In particular, a general-interest journal, as opposed to a specialized journal, cannot carry only weighty contributions to science, it would sink. It needs lighter material too, it needs to combine being authoritative, informative, sometimes irreverent and often entertaining.

The editor must steer the journal with an eye to maintaining and if possible improving quality. This is gauged in part (in part only) by the impact factor based on citation analysis; also by peer survey and specialist opinion. The editor is responsible for a high profile scientific image, but not at the cost of the journal becoming widely unread, a mere archival repository of articles.

Thus the editor has to keep in mind a diverse readership. Yes, the editor is responsible as much to the readership as to the authorship! This readership ranges all the way from impatient young readers fluent in IT (information technology), who read preprints online and are often interested only in a narrow specialized topic, to the mature reader for whom the journal is more like a newspaper – carrying news, background articles and editorials.

An initial task is among the most important of all your functions. This is appointing your editorial staff. The editor-in-chief must be surrounded by equally competent associate editors. Finding those people and convincing them to assume the job is difficult.

While the editor has to be ecumenical in the choice of topics to be featured, but still impartial when arbitrating between personal tastes and those of the colleagues, between various sub-editors each pushing for a particular subdiscipline, between authors and referees. Combine open-mindedness with authority. In some cases, the editor of a journal needs to override the recommendations from the referees.

Your skills as a negotiator and as a diplomat will often be required. The editor must negotiate with the publisher for page limitations, for changes in format, for technology upgrades; with the other editors for changes in direction, for guidelines in policy, for implementation of new technology, for demanding higher quality. You will need to scrutinize ethical issues, both in principle and as a matter of conscience, and in some cases for practical reasons (avoiding legal action being taken). These negotiations must be carried out in such a way that the editor can keep the editorial ship moving without being autocratic.

Another quality or talent has to be mentioned. It is an essential ingredient. A fine journal editor needs to constantly project into the future, asking questions about the state of the field in two, five or ten years, the technology of publishing at such times (unless he can draw on expertise in that area), the profiles of the journal and of its readership at those intervals.

Yet more important is intuition, allowing the editor to smoke out the original papers, those of likely lasting value because they innovate and are radical departures, amongst the incoming flow of manuscripts.

The highest reward you will have as an editor is the privilege of holding your finger on the pulse of science in the making. Exercise it with total independence, considering yourself fully responsible for the whole editorial content of your journal.

REFERENCE: M. Alley (2000) *The craft of editing: a guide for managers, scientists and engineers.* Springer, New York.

stylish and careful writing 

EDITORIAL

The analogy is with an exquisitely-wrought piece of jewelry, a watch, whose interlocking parts mesh smoothly. It is a text which demands stylish and careful writing.

It belongs within the genre of the essay, but it is an essay in miniature. It thus addresses an issue and suggests a way out, a solution.

The issue is typically at the interface of public interest and scientific findings. Examples of such current issues include genetically-modified organisms, human clones, and global warming to mention a few examples.

State your position about the issue you have chosen to address. Muster scientific arguments which bolster your position. Do your homework carefully as a preparation. Have all the facts at your fingertips, those which agree with your stand as well as those which can be interpreted as being against your position.

By its very nature, an editotial turns a personal into an official position. You write it either as an editor or as a guest editor in a newspaper, magazine or a journal. Sometimes, it is signed and sometimes it is anonymous.

Short in length, devoid of ornaments, the editorial requires quality writing and ruthless editing. Make every single word meaningful. Make sure that it is published in the most appropriate place.

REFERENCE: M. Weingarten (2005) *The gang that wouldn't write straight: Wolfe, Thompson, Didion and the New Journalism Revolution.* Crown, Random House, New York NY.

just like cream 

INFORMAL DISCUSSION

When and where does it happen? Whenever and wherever scientists meet, in person or in the ether. Greeting a colleague at the airport, who comes to give a **seminar**. Making a **phone call**. During conferences, outside the lecture hall, in-between sessions or during meals. Traveling to a conference, sharing adjoining seats with a colleague on an airplane or a bus.

What, then, are topics suitable for discussion? They range widely. Gossip is foremost: who among the people we know received a grant, got promoted or is about to move to another institution. Positions available. Sources of money, a foundation whose existence we discover.

In addition to this familiar exchange of professional news, circulating on the grapevine, many other types of information are being traded: suppliers of instrumentation or chemicals; useful websites; a key reference one had overlooked in a **bibliography**. I also rank foremost in such verbal exchanges the singular piece of know-how: it is perhaps crucial to success of a procedure, it was not made explicit in the experimental part of the published work. In brief, such off-the-cuff conversations are the very blood of scientific life, they are what glues the community together.

Since such discussions can occur at any time, usually without advance planning, always carry a recording device, such as a pen and a small notepad, or a few filecards. Better yet, a commonplace book is handy. I have always carried one in my hand throughout my career, for a total of about 60. I numbered them in sequence on the cover and on the spine. I also numbered the pages, prior to compiling a table of contents when I had filled it up.

Transfer of the information garnered is as important as acquiring it; You need a central repository, whether the just mentioned commonplace book or your laptop. Make it a habit to do so as soon as possible after the encounter, while the conversation is still fresh in your mind. You will have a better chance at remembering the infospeck for having seen it twice. It will enable you to transcribe your initial jottings into understandable, fully-formed and logically ordered sentences. It will also allow you some indexation, to help future retrieval.

Another good habit – skirt laziness there – is to address a follow-up memo to the person you have met, in order to make clear who said what. It may come in useful later on in any kind of argument, about priority for instance.

Likewise, be a good sport in giving credit to whom it is due. When referring, in print or in speech, to a piece of information, make sure to refer to this particular conversation, to where and when it took place.

Most important in such episodes though, is not the information exchanged. It is the buildup of personal relationships. The scientific community is essentially the sum total of such informal links. It is a network of buddies. Hence, during the encounter, communicate who you are. If indeed you are a pleasant person to interact with, one who also has something to say and to contribute, someone who is able to take an interest in work outside your own boundaries, you will rise to the top – just like cream.

REFERENCE: E. Goffman (1959) *The presentation of self in everyday life.* Doubleday, Garden City, NY, (numerous subsequent reprints, Penguin Books, Harmondsworth, Middlesex, UK).

documentary resources 

INTERNET

This is common knowledge, Internet and the World Wide Web were imagined and devised by some of our fellow-scientists. The story started in the early nineteen-sixties. J. C. R. Licklider of MIT wrote a series of memos in August 1962, to present his, then futuristic idea, of a globally interconnected set of computers. An organism funded by the Pentagon, DARPA, gathering scientists engaged in defense-related research started developing in the late nineteen-sixties ARPANET, the predecessor to the Net.

The first public demonstration of the new network to the public occurred in October 1972 at a conference in computer technology. The same year, e-mail was first introduced. The technology enabling transport and forwarding of information, which goes by the acronym of TCP/IP, began to be implemented in the 1970s. Another crucial piece of technology, the ETHERNET, was devised by Bob Metcalfe at Xerox PARC in 1973.

By 1994, when the then vice-president of the United States, Al Gore, lent the prestige of his office to the phrase »information superhighway«, the Internet had become a reality for almost everyone – at least in the most industrialized countries. By now, the figures regarding the amount of information stored or exchanged are colossal.

The Internet thus offers awesome documentary resources to a scientist. These are too well known to reiterate here. However, make sure to carefully crosscheck the information gleaned from the Web. Systematically use several independent sources and if possible, try to locate a similar piece of information in a library journal. Also, quote the url when you cite information obtained from the Web. Respect the laws, both the moral law against plagiarism, and the intellectual property laws regarding copyright.

The Internet owes its success to being a global network without any central control based on a policy of absolute freedom in dissemination of information. This was the case at least initially. Governments have now stepped in and begun to interfere. The principle of total freedom was reasserted in 1996, when the US Supreme Court ruled that the Internet is »an endless world conversation.«

The rules of good conduct on the Net are probably already familiar to you. I shall thus content myself with a reminder. Do not contribute to the verbiage, do not send unsolicited material to all the recipients on your mailing list, you risk antagonizing a few. This includes political activism. Ask yourself whether you would not be more effective writing a letter by longhand to half-a-dozen carefully selected recipients. Respect privacy. For instance, do not divulge, inadvertently or deliberately, the e-mail address of a colleague without receiving permission to do so.

Unfortunately, a majority of cybernauts espouse an antiscience stand. Be aware of it. Let your skin thicken, if need be. Avoid the chatrooms, the discussion groups. In my experience, the signal-to-noise ratio in these locations is very low. Often, information of doubtful value circulates there. Furthermore, a few people tend to be abusive under cover of their chosen nicknames.

REFERENCE: J. Kissel (1996) *Cyberdog: Opendoc on the Internet.* Wiley, Hoboken NJ.

rely on your wits 

KEYNOTE LECTURE

You have been asked, and you have accepted to deliver the keynote lecture in a conference. This raises issues of representativeness, of selection of the results to be shown and commented upon, of the oral versus textual, and how best be influential. As a first step, solicit a maximum amount of information from the organizers. Be warned it won't be much. You will have to rely on your wits to design and put together that **lecture**.

First, the issue of representativeness: the lecture cannot be restricted to your own work, sensu stricto. Attempt to map out, and to wrap into what should be a unified whole, the sometimes conflicting goals of the organizers, the participants and yourself. Needless to say, but let me nevertheless remind you: selecting slides and/or transparencies which you already have in stock for your standard lectures fails to heed the call. You need to put together something different. It will be an innovative talk for you to present.

How then should you select the results, predominantly yours of course, also from others (you will need to ask for permission, if only informally), which you will comment on? You have to take the long-term perspective. You need to visualize the conference as a link within a continuity. Prepare your contribution accordingly. Show a keen sense of history. You need to take stock of a whole field, of where it has come from and of where it might be headed.

To continue to deal with this first issue of representativeness, for the one hour you will have to fill, your lecture should symbolize the state of the art in a discipline or sub-discipline. Your persona and your stature will be, for that short time, those of a statesman of science. Above all remain simple and modest. Do not become dogmatic and pompous. Do not hesitate to use irony, and to indicate subtly that you are not allowing yourself to be put on a pedestal.

To return to the issue of data selection, choose results from your work and that of others not only for their elegance, cogency and being meaningful even in an allusive mode; remember too their insertion in a more or less linear argument: there has to be a smooth flow, whether the strand you will follow is one of logic or one of chronology.

Oral or textual? Memorize your text well enough that you won't be reading notes. It will be carried along by a splendid iconography, which you have picked with extreme care. However, for maximum impact, building on the urge from some of the attendees who will need to refresh their memory (not everyone takes notes), insist on publication of your comments.

Offered an opportunity for telling your peers where they stand, is a task not to be confused with passing judgment, but one to be carried out with a combination of lucidity, articulateness, and vision. It will set the stage for the conference that is to follow. It will also make you stand as a figurehead, and the envy of some, remain aware of this. If you do it lightly and with **wit**, you will discharge yourself honorably in what is a rather delicate task (and quite an honor).

REFERENCE: M. Alley (2003) *The craft of scientific presentations*. Springer, New York.

tell a tale

LECTURE

To prepare a lecture and to deliver it amount to revisiting a landscape, already familiar for having been explored and conquered. The lecture covers part of one's work. To be a scientist is also to be a teacher. The notion of the scientist-teacher dates back to the early nineteenth century when the brothers Humboldt devised the first modern university in Berlin. It makes much sense. One talks well of what one knows from having studied at firsthand, perhaps even from having discovered it. Conversely, teaching is the best way of finding new avenues of research. They stand out as gaping holes in the available knowledge, when we present it in teaching.

To dissociate the scientist from the teacher is no more feasible than to separate the two sides of a coin. Outstanding organizations exist for pure research, such as the French CNRS. They lack the teaching dimension, to their detriment. Research at the forefront demands to be shared. The lecture is, together with the publication, the best channel for such sharing.

Let us then assume that you are about to lecture about your work. Give your lecture a structure. the problem-solving mode is perhaps the easiest. State first the problem to be solved. Then present your solution. Close by giving the solution yet more general scope.

In any case, whatever the structure chosen, to lecture is to tell a tale. Your narrative will be all the more enticing that you will get your audience – of peers, most likely – to identify with you. Hence, you need to convince them that the problem you had chosen to attack was important. You will explain why solving it was difficult. You will continue your narrative with how you got the idea of a solution, and how you put it into practice. You will mention the obstacles which had to be vanquished. And you will bring your story to its presumably successful conclusion.

Such storytelling is also termed »to spin a yarn«. Could it be that women were the first to recount key events in speech, thus serving as the living memory for the tribe, while preparing textile fibers to be woven into fabrics? The conjecture is highly plausible. Set yourself a comparable goal. Make your lecture memorable to your community of peers.

Why is the lecture the main tool of teaching, superseding the **textbook**, the reading of notes and exercise solving? Because there is this flesh-and-bones person, on which converge the eyes of the assistance, who combines storytelling and explanations. This person remains acutely aware of whether the audience is following, or not. Which enables you, as the teacher-lecturer to backtrack, to try out a different exposition or explanation, until everyone has grasped the point.

Which brings up the duration of the lecture. Do not overstay your welcome. Do not speak for more than about 50 minutes. Let people know, periodically, how much more time they will have to listen to you. Moreover, be aware that the attention span of the listeners is no more than about 5–10 minutes in a stretch: segment your lecture accordingly. Rekindle interest in your listeners half-a-dozen times during your presentation.

How? By switching from a topic to the following one. By alternating between the heavy stuff and lighter fare, such as demonstrations, anecdotes and jokes. By engaging your audience into the argument, by putting questions to them. A lecture is a collective construct. To reduce it to an active speaker and a bunch of passive listeners is a beginner's mistake. Make the audience participate. Your lecture, in so doing, will be much improved. This is the main role of anecdotes.

Remember the Golden Rule for giving a lecture. Start by telling people what you are about to say. Then say it. Finish by telling people what they have just heard. In other words, do not forget to recapitulate the important points made during the lecture.

The Golden Rule has a complement, which I shall term the Platinum Rule. Start by telling people all the good news they want (and expect) to hear. Then, tell them all the bad news they did not come nor expect to hear. Finish by telling them the take-home lesson from such a mix.

The Platinum Rule expresses the tension in scientific life between the consensus we strive to build (»the good news«), and dissent which is even more critical to science. Any discipline which slides into complacency is in mortal danger. Group think is antinomic to science. The role of the intellectual is to challenge the stultifying consensus.

A lecture has thus to fulfill the two complementary functions, of presenting new knowledge in an easy to grasp and engaging manner, and of showing how this new knowledge cracks the veneer of the established orthodoxies. Do so without hype, without one-upmanship, without raising your voice, in an unassuming and modest tone. You are only the bearer of the news.

This attitude, while it will go over very well with your audiences, may create some trouble for you at home. We scientists are so much used to rational thought, and to the verbal idiom for presenting our argument and conclusions, that we tend to use the identical tools for thought at various junctures in everyday life. This often makes for a sententious tone. It can all too easily be mistaken for our claiming to be a Mr. or a Ms. Know-It-All. »Do not lecture me« is the attendant standard complaint of our partners. Not only won't they listen to us, they will give more credence to the mere opinions of any Tom, Dick or Harry, or their female counterparts. however well informed and well-argued your reasoning, it suffers from its ponderous utterance. Small talk from Tom, Dick or Harry will easily win the day.

Your lecture needs not only balance between the good news and the bad news, it also needs to be balanced between the spontaneity of someone regaling friends with a good story and more formal, better polished discourse. I have always found it very useful to write out and memorize both my opening and my closing **sentences**, in order to fulfill the latter need. This little trick always works very nicely.

REFERENCE: R. L. Weaver (1982) *Effective lecturing techniques: alternatives to classroom boredom.* New Directions in Teaching 7:31–39.

the oratory superb 

LECTURE SERIES

You have achieved celebrity status. You owe it to your research achievements, to the books you have published, above all to your conquest of the media; yours has become a household name. Accordingly, you have been invited to give a series of lectures in a great, renowned university. You wish to be worthy of the honor, you have been called upon to address a general audience.

If a scientist, it is much more likely that you were asked to contribute talks in your area of specialization. Is that narrow-minded? Somewhat. As one of my correspondents comments on that point, »*I wish more lectureships were like these, pushing the recipient to speak to a general audience. Most science departments solicit the money for* lecture *series, and then having in hand the carrot, the honor and honorarium, fail to do a service to the community, and fail politically to build bridges outside, by not pushing for at least one general lecture*«.

Hence, you need to take the time away from everything else and polish your lectures. These series often go in threes. First, you have to choose the subject matter and to decide on a general **title**. You also need attractive titles for the individual lectures.

Remember, you will address a general audience, not one of specialists. Therefore, your topic has to be rather wide-ranging, a trans-disciplinary catch-all. Colleagues in your discipline will make up at most perhaps 10% of the audience. This differs considerably from what you are used to when giving lectures. Besides a rather general topic, the approach will have to be innovative and original, the **scholarship** impeccable, and the oratory superb – in a word, you will be expected to shine. It will be a good idea to arrive at University X fully prepared, and to have finished toiling over the *written text* of your lectures, before you deliver the first one.

The reason is that, as soon as you step onto the campus of University X, you won't belong to yourself any longer. True, you will receive a hefty honorarium for presenting this series of public lectures. In exchange, you will be indentured to University X for the duration of your time there. You will be tied to a heavy social schedule. You will be called upon for formal meetings, luncheons, dinners with the President of the University, with Deans, meetings with students, with colleagues in your discipline, with the family of the donor of your named lecture endowment. If, as is entirely legitimate, you would like to preserve some time to be by yourself, for something as mundane as doing a little work, you will need to make explicit arrangements well in advance. A good piece of advice is to schedule an appointment after lunch each day that you have with a faculty member who seems to exist on each campus, Dr. Knapp.

Another reason for preparing a script for this lecture series is that you are likely to be called upon to publish it, either with the University Press or with a commercial publisher having entered into such an arrangement with University X. Do not content yourself with a selection of slides, transparencies or PowerPoint slides. This is not enough. You will have to uphold your reputation by reaching for the next level up, namely a series of well-written, polished lectures.

Reflect a little on your invitation. It amounts to a trading agreement, within an economy of status. University X is benefiting from your reputation, but its own prestige reflects favorably on you too. Those are the terms of the exchange. For University X, your name goes into a long line of at least equally prestigious scientists and scholars who have previously illustrated this particular lecture series, be it the George Gamow Lecture at the University of Colorado, the Leverhulme Lectures at Oxford University, the Trevelyan Lectures at the University of Cambridge, the Ira Kukin Lectures at Yeshiva University, the Louis Clark Vanuxem, the Scribner or the Stafford Little lecture series at Princeton University, etc. There is also the possibility of University X looking you over prior to hiring you, or trying to do so.

But let us leave the invitation and its interpretation and return to the actual lectures, to bring this to a close with a short list of pros and cons. You will be in the limelight, you will receive lots of money, and you will meet interesting people. The down side is that the experience will be exhausting, whether for just a few days or several weeks, you will be plagued at time with the company of utter bores, attendance at your lectures may drop considerably between the first and the second, which is no reflection on the quality of your performance. And do not forget this final admonition, brilliance comes at the cost of depth. Finding the proper balance will be a challenge.

REFERENCE: S. E. Lucas (1983) *The Art of public speaking.* Random House, New York.

your unique voice 

LETTER FOR PUBLICATION / OP-ED PIECE

Why include this seemingly minor genre? Because the voice of scientists is all too rare in the media, restricted as we are to giving our opinion when asked, as with an interview.

Use your unique voice. You won't have to seek the occasion, it will impose itself when you come across a truly inflammatory statement: stating a patent untruth, ignoring the record, or just plain stupid.

Before turning to your word processor to correct the record, decrease your blood pressure. Your main goal is publication. Anything beyond a single, short **paragraph** won't get published, or will be heavily edited. By contrast to such an unsolicited letter, a commissioned Op-Ed piece would be printed at the length it was specified for. Be brief. Work hard on the key **sentence**, which may possibly be your punch line.

It is a difficult text. You need, in 100–150 words maximum, to show yourself as scholarly, authoritative and witty. Your aim is to correct disinformation. Most probably, your letter has to take an unpopular standpoint. Thus, you risk preaching in the wilderness.

You need to puncture a stereotype. Doing so attacks conventional wisdom. The goal is to reform collective attitudes, perhaps even in current fashion, emanating from group think.

Group think – an oxymoron if there was one, i.e., a contradiction in terms – is otiose and lazy. It always shelters beneath a prejudice. It amounts to a cold version of collective hysteria. It compares in lack of lucidity and weakness of expression to a report written by a committee.

Conversely, it feels very comfortable to be part of a group. The intellectual life, however, achievements of a scholar, or those of a scientist, all thrive on individuality. To secede from group think, to lambast it as group unthink, is an ethical choice too. As wrote Henri Michaux, »*he who sings in a group, if called upon, will throw his own brother into prison*«.

In drafting your letter or an op-ed piece, keep in mind that if published, it will be archived and become fodder to historians, a most legitimate and desirable goal. Hence, make sure that you know the facts, all the relevant facts and that you state them accurately. If you make any mistake, however puny, someone is sure to pick it up, and to use it to destroy your credibility.

Refrain from the »on the other hand:« too many letters to the editor start by being obsequious to the journalist who wrote a piece, and term it admirable, prior to attacking it (»I take issue with …«, »I agree that X, but …«, »I cannot dispute Y but …«). Readers and editors alike are interested in your own voice.

Skip the polite formulas, go for the issue, straighten it and sign off.

REFERENCE: for the opposite viewpoint, see the well-argued and eloquent defense of group think in J. Surowiecki (2004) *The wisdom of crowds: why the many are smarter than the few and how collective wisdom shapes business, economies, societies and nations.* Doubleday Books, Random House, New York NY.

a mission-oriented stance 

MAGAZINE ARTICLE

The magazine article is arguably the foremost medium for science communication for the general public. In recent years, Stephen Jay

Gould's columns in *Natural History*, articles in *The New Yorker* by Oliver Sacks (and a few others), those in *American Scientist*, in *The National Geographic*, in *The Smithsonian*, those by Sharon Begley in *Newsweek* before she went to work for *The Wall Street Journal*, essays in literary reviews such as *The American Scholar* or *The Georgia Review*, all have nourished the genre. Even though some are illustrated, magazine articles succeed (or more rarely, fail) on the strength of the text alone.

How then should they be written? Steer clear of the **research report**. What you need is almost the exact opposite. When you write a research report, or its progenitor, the **research proposal**, you assume a mission-oriented stance. You spell out the goals, the means for reaching them and what has been accomplished so far. In other words, you describe how you propose to go from A to B, or how you have done so. You make it look straightforward.

The magazine article conversely takes its readers for a stroll. Not from A to B, nor from L to M, but on more fanciful itineraries, such as L → D → O →F →M. Such a wandering path somehow recaptures the charm inherent in natural history or in the voyages of discovery of yesteryear.

The magazine article most often is a personal essay. It takes as its implicit model the belle-lettrist essay, the form which Michel de Montaigne made famous in the sixteenth century and introduced to the canon of Western culture. This is the literary genre which has been illustrated, in our time, by the masterpieces, to cite just a handfull of writers, some long gone, and some still very much with us, Peter Medawar, Lewis Thomas, Loren Eiseley, Henry Petroski, Roald Hoffmann, and Philip Ball.

While a magazine article may give an impression of nonchalance, visiting a strange and lovely landscape, exuding a sense of wonder at its intricate beauty, you will want to plan a detailed itinerary, to be masked or erased from the view of the reader once you are done. Apparent fantasy demands extreme rigor to buttress it.

The distinguishing mark of a magazine article on a science topic is for it to combine scientific rigor and accuracy with personal introspection and freedom. You need to be a stickler for the accuracy of the scientific facts you cite. Likewise, your reasoning must be logical and upheld by the evidence. But your whole approach is the apparently random walk, transdisciplinary if need be.

And idiosyncratic. The magazine article is a delight to read if (and only if) your personality shines through. It is, truly, an excellent medium for communication. You will touch many readers with it. You will be surprised at the explicit response you will get from quite a few of them, some of whom will send you a note. Is there better evidence of having struck a chord?

Which brings up my last recommendation: writing a magazine article should give you the feeling of writing a letter to a friend. However, they are distant friends, hence your style cannot be totally informal. You have to remain somewhat formal and distinguished. Nevertheless, your disquisitions need to be couched as a narrative, one which will let the reader hear your genuine voice when you recount episodes from scientific life.

REFERENCE: P. P. Jacobi (1997) The Magazine Article: *How to think it, plan it, write it*. Indiana University Press, Bloomington IN.

pithy, descriptive and general 

MONOGRAPH

To write such a book is a sign of both scientific maturity and acquired mastery of one's field. Otherwise, do not make the attempt, it is not worth it. Why do it, unless you are building a monument. You have to create a classic. Such a text combines knowledge and wisdom. You may have reached such a point in your thirties, more likely you are in your fifties or beyond.

How does one know that the time has come for such a venture? By your often being invited to give **keynote lectures** and/or **lecture series**. Because of an acquired lucidity: you are able to see clearly both the statics and the dynamics in your field, i.e., its inner architecture and where it originated and is now heading.

Why do it? To express your expertise in a lasting manner. To imprint the field with your name. To influence young scientists. To be read by nonspecialists, drawn in by the combination of expertise, scholarship and good writing. Because of your responsibility, not only to history, to yourself as well.

And how does one find the time to write it? You are already overextended, with a more-than-busy schedule. There are two solutions to the predicament. You can write your masterwork before your workday, by making a decision to get-up a couple of hours earlier than usual, devoting that daily period to putting together this book. Or you take advantage of a sabbatical to get started. If, when it is over, you failed to write more than a third of your *magnum opus*, things do not look good. You'd better drop it.

Assume then you have made the decision to do it. Coming-up with a title will be easy. It has to be pithy, descriptive and general. Avoid being cute. Go against the present vogue of subtitles, it won't last for ever. Take example on some classics: *Symmetry, The Principles of Nuclear Magnetism, Liquid Crystals, Quantum Mechanics, Valence, …*

Your next step is **organizing your material**. This is where your bird's eye view of the field, your understanding of its organic structure will make it a natural, a sweeping gesture. Your chapters will correspond to the lines of forces in the field, starting with the foundations and with the basics.

Once you have done so and put together an analytic table of contents, it only remains to flesh it out in writing. Remember that you are dealing with the whole field in an objective, dispassionate and superior manner. Hence, avoid focusing on your own contributions. You will treat them on an equal footing with those from your colleagues, friends or foes.

One of your traps, the deepest, is to fall in for comprehensiveness. You do not want to write a lengthy treatise, it will only gather dust on library shelves. Brevity is golden, here more than almost everywhere else. Take example again on such classics as Dirac's or Abragam's. Write concisely.

Each sentence should burst with meaning and yet be structured with utter simplicity. Each paragraph will condense years of experience, at least a dozen publications (to mention an order of magnitude). Each chapter should connect with its neighbors in an harmonious way. Throughout, write an elegant, a fluid prose.

There is no better way for optimizing the product than submitting each completed chapter to three or four colleagues, whom you trust for their competence. They will spur you to add on to the edifice. Resist the impulse. Most of the time, you will be able to include their input in the notes rather than in the body of the text.

Choosing a publisher, once the manuscript is under way and progressing at a good clip, will be easy. Far easier than turning down a publisher pressuring you to write a monograph you are not truly ready yet to produce. Go for a university press of the first rank. When you have received two or three offers for a contract, go for whichever publisher gives you both free rein in the design of the monograph, without any commercial consideration and a highly professional editing.

They will grace your book with a handsome layout and production. They are even more interested than you are in being able to boast of such a gem, of lasting value, in their list.

REFERENCE: J. M. Williams (1990) *Style: toward clarity and grace.* University of Chicago Press, Chicago, London.

warts and all 

OBITUARY

Strict rules apply to such a set piece. Besides birth and death dates, circumstances of the demise, it also includes early childhood influences; the studies and, important, who the mentor was; a summary of the main scientific contributions; mention of the family; a **paragraph** on the hobbies of the deceased; and, if possible, a revealing anecdote.

The British have turned this morbid genre into an art form. Their daily newspapers carry obituaries which are little gems of writing. There are even collections in book form, such volumes of *Obituaries from the Times* are prized possessions on my bookshelves, I consult them regularly.

The revealing anecdote? Here is an example. After Stephen Jay Gould's obituary appeared in *The Guardian* (May 22, 2002), Alan Andrews wrote in (May 31 issue) a small correction. He added the following recollection: »*Steve incandescently threw himself into what was already a very vigorous student life at Leeds. …(He organized) weekly demonstrations outside a dancehall in Bradford which refused to admit black patrons; the demonstrations continued until the dancehall was integrated.*« Such anecdotal material is precious for revealing the person beneath the lab coat.

Early influences? A well-crafted sentence may suffice, as in this obituary (*The Guardian*, October 26, 2004) of the physicist Sebastian Pease, written by Robert Hinde and Hoseph Totblat: »*His scientific expertise and social conscience echoed his family background, with its Quaker roots*«.

To mention hobbies is charming, they often reveal an unexpected side of a personality. Harry Pugh wrote in (*The Guardian*, October 23, 2003) to add to the published obituary of the scientist William Hall (October 8, 2003) that he »*was a superb exponent of home brewing*«. The obituary itself, written by M. J. Harris, described Hall's personality with the carefully-chosen terms, »*approachability, insight and encouragement*«. Lest one should think that it failed to do justice to Hall's hobbies, he, the first holder of a chair in nuclear engineering at Manchester University, had an »*interest in, and enthusiasm for music*« and »*also had a passion for building model steam engines*«.

Some obituary writers use features of their topic to express views they hold important. I am personally in favor of selecting a few such stepping stones. For instance, Steven Rose's obituary of Stephen Jay Gould (*The Guardian*, May 22, 2002) makes a number of sweeping generalizations: »*The intellectual's development from radical young Turk to mature senior academic is traditionally that from iconoclasm to conventional wisdom – cutting-edge researchers are often ignorant of their own science's history – the sort of whiggish, anecdotal approach by which senior scientists tend to ossify the progression from past obscurity to present clarity.*«

There remains a crucial issue for the obituary writer: to tell everything, warts and all, or not? Two considerations support restraint and keeping one's peace: an obituary is supposed to sum up the achievements of a person, not trash him once dead and unable to defend himself; there is also the necessary respect for the survivors in the family: *de mortuis nil nisi bonum*, of the dead let nothing bad be said.

Our duty to the truth stands conversely in favor of full disclosure. Too often, obituaries suffer from their hagiographical tone and the sense of the real person is lost. These issues came to a head recently. The *British Medical Journal* published in May 2003 a scathing obituary of the scientist-cum-entrepreneur David Horrobin, attacking him as a »*rotter, a snake oil salesman, a chancer*«, which did not go unnoticed. Should you call a spade a spade and a scoundrel a scoundrel? I leave it to you (and your lawyers) to decide.

REFERENCE: B. Kovach and T. Rosenstiel (2001) *The elements of journalism: what newspeople should know and the public should expect.* Three Rivers Press, Crown, Random House, New York NY.

anticipating the likely issues 

PANEL, ROUNDTABLE DISCUSSION

A successful panel discussion resembles chamber music, harmonious, functioning as a whole, making a useful statement and thus being pleasurable. The panel consists of a chairperson and a small number of members, for a total of six or seven, preferably no more.

The role of the chair differs markedly from that of the members. He or she is responsible for the quality of the discussion. Called upon to chair, prepare your opening statement very carefully: about 12 **sentences** at most. Summarize the issues. List the points to be discussed in sequence.

Briefly introduce each of the panelists. It will then be their turn to make their own opening statements, announcing their views and launching the discussion proper. As the chairperson, it is your responsibility to cut short long-winded interventions, to puncture hollow statements and to correct mistaken ones, if any. Bring order to the discussion. Take up the successive points in sequence that you have planned for. Prevent attempts from any among the panelists to take over and turn the roundtable into a pulpit. Make sure that everyone expresses views equitably, timewise.

If you are a panelist, prepare for the discussion by anticipating the likely issues. Jot down your view of each. If you have a minority stand or an original take on the issue, prepare a statement in writing. Prepare it to be delivered in a minute or two maximum. Ideally, this articulate viewpoint should highlight you as both a thinker and a talented public speaker. Bring a paper pad, pen, cue cards (crucial) so that you will be able to quote any data accurately. Make sure to have a glass of water and throat lozenges nearby. Respect the authority of the chairperson. Avoid one-upmanship. Do not even attempt to upstage anyone else on the panel.

Practice always helps. You will find that, by rehearsing the format with your coworkers, or from the cumulative experience of taking part repeatedly in such discussions, that you will get better at it: presenting your views clearly and concisely; summarizing in a sentence the whole argument you are about to rebut.

On a panel, your most delicate task as a member is to get the floor when you have something to say. You have two options. Either a discreet gesticulation to catch the eye of the chairperson at the right instant, a split second before it is time for you to intervene. Or taking advantage of a silence, or a breathing spell from a fellow-panelist, to contribute. Whatever the case, speak in a natural way. Make sure to give the audience silent spaces for your message to sink in.

I have so far failed to mention a potentially devastating ingredient – the audience. Failing to address questions from members of the audience is discourteous. Responding to questions too early may derail the whole proceedings. If you serve as the chairperson, it is your responsibility to trigger such questions and comments, by specifying clearly the point to be addressed and soliciting questions from the audience with direct bearing on that point. Make sure that the person about to ask a question stands up and identifies himself or herself.

The concluding statement should be concise, indicating the points of disagreement and spelling out the consensus reached, if any. This will be the take-home lesson for everyone. You need to encapsulate the proceedings in a fair, even-handed manner, without letting your own views predominate. Think of yourself as a spokesperson who gives the gist of the debate.

REFERENCE: S. E. Lucas (1983) *The art of public speaking.* Random House, New York.

a commercial transaction 

PHONE CALL

Remain aware of what people tend to forget: a telephone conversation between two scientists amounts to a commercial transaction since information is being exchanged. Whether you initiate the call or you receive it always have paper pad and a pen at hand. During the course of the conversation, jot down the key points made by each side.

Besides sharing information, a phone call is also a social binder within the scientific community. It is basic, together with **e-mail** and conferences to the grapevine. Gossip travels along such channels. For this reason when you pick-up the phone to talk to a colleague, or vice-versa, do not bypass the small talk. It may be the whole purpose of the call, or it may be an apparently minor part of it. In any case, it is basic.

Be very aware of the personal characteristics of the other speaker. Distrust so-called operators and keep them at arm's length. They talk to you in order to obtain information for their own use, without giving much in return. Trade only with fellow-scientists whom you trust to be colleagues with a fine reputation, whom you respect for good cause.

At the very beginning of the dialog, be prepared to state the scope, the points to be discussed and the intended duration. This will help you, when you utter the standard excuse to end the conversation: »*forgive me, I must leave in a couple of minutes to attend a meeting, that I mentioned earlier«*. This will help you to retain a measure of control, both over use of your time, and therefore over the whole proceedings.

In any case, after putting down the telephone, back-up the conversation you have just had with an immediate follow-up, by e-mail for instance. This will recapitulate the main points discussed. It will serve as a pseudolegal protection against the information you gave away being appropriated without your consent and without proper credit to you.

REFERENCE: P. H. Andrews (1985) *Basic public speaking.* Harper and Row, New York.

telling a story 

POSTER

The poster is a relatively new way in which to present advances in science. True, its ancestry goes back several centuries, to the Renaissance. Emblems became in vogue in the sixteenth century. An emblem had three parts: one was a picture. The second was a textual commentary, underneath the image. Often, the picture had some enigmatic or surrealistic component, which the commentary drew attention to. The third part was a motto.

You have undoubtedly recognized remnants of Renaissance emblems in billboards and commercials. Closer to us, the poster originated with conferences in molecular biology, when this then new field was expanding vigorously in the nineteen-sixties and early nineteen-seventies. Organizers of such meetings had to cope with numerous participants – too many. There was a surfeit of papers. Yet, authors had to present a paper in order to become eligible for financing for the trip. Poster sessions were a means to obviate this conundrum.

Turning now to the main features of a poster – needless to say, I am describing what it ought to include, ideally – this communication channel is a means for telling a story. People who are passing-by your poster, and who might be drawn to it, will make a quick decision as to whether to give a closer look, perhaps even find out from speaking to you what it is about. Hence, you need to make the story you are telling on your poster vivid and eye-catching. You will aim for distinction, originality and elegance. Your poster needs to be picked out from a whole crowd.

A poster ought to be visible at a glance, easily readable from a distance of six feet or so. It needs to tell its story, unaided: additional information provided by the author, once approached, will flesh out details for those viewers with special interest in a detail. You need to organize the flow of information for the eye, linearly, from left to right and from top to bottom (forgive my stating the obvious, but something like 85% of the existing posters flaunt this particular rule). Make sure, in addition, that your poster carries a hook, some element with which to seduce the would-be client; that the pictures it shows are both vivid and pertinent; and to strike a balance between images and text.

How to go about it? Start by defining to yourself your main message. Write it down. Can you make it less than about 25 words? In that case, your poster is already half-done.

To start improving your first draft, box the copy. As the next step, cue the eye in some way to the sequence in which those boxes ought to be read. This will immediately improve markedly the quality of your poster.

Now to the design. Wrap your poster in a visually attractive idea. In other words, find a way in which to present the gist of your results, to make them leap to the eye. Devise a way that is astute, striking and distinctive; and which at the same time is pleasing to the eye, because of the harmony, of the type, of the colors, of its inner rythm, ...

The most frequent mistake is the posting of a reprint, in which case you advertise to the world your laziness and your lack of imagination. A second, very widespread error is using the whole available space. Editing your poster to occupy only a fraction of the allotted space, in itself, designates your poster as special. A third pitfall is being dull – advertising is allergic to the bland and the uninteresting.

Many a poster has self-defeating defects, such as excessive use of the passive voice (»methods described in this poster«, »methods intended to be practical«, »applications to x, y, z are shown«), likewise abuse of **acronyms** (do you really want to communicate only with specialists of your own subfield, don't you want to draw in a larger number of people?), an encoded and nonobvious use of colors, or likewise too loud screams from words underlined, in italics or in bold for added emphasis (use each such ploy sparingly).

Remember foremost that this is a piece of advertising. You use it to present a single result, idea or discovery (emphasis on SINGLE). A poster is an announcement. It needs good copy.

Thus, it should read like headlines in a newspaper, not like material in a textbook. In addition, a poster session at a conference is a social occasion. You will meet with fellow-workers from the time you were a graduate student or a postdoc. Accordingly, the poster is very much the equivalent of a quick chat with someone during a cocktail party: it shows your best scientific face to the world, it encapsulates your work in a memorable manner.

REFERENCE: J. Franklin (1994) *Writing for story: craft secrets of dramatic nonfiction.* Plume, Penguin, New York NY.

Respect and friendship 

PRESENTATION TO A VISITOR

Comradeship extended to a fellow-scientist is your main message. It determines the worth of your pitch. When your visitor shows up, give him or her the warm feeling of friendship. A comfortable armchair, the offer of a glass of water or a cup of coffee, directions to the nearest rest room, an offer to lend your computer for checking the **e-mail**, physical closeness – attention to such items works toward a feeling of intimacy. If there is to be a meeting of minds, without which there won't be any flow of information, it is crucial to take care of such physical details for starters. Respect and friendship are your key assets.

Keep your talk short, within 15–20 min even though the meeting may last for an hour – in which case conversation on general topics is called for after you have presented your research. Writing on a note pad or on a blackboard are far preferable to using already existing **visuals**. The latter is tainted with a practicality which goes counter to your aim, making your guest feel special. Do not insult his/her intelligence, explain only what needs explaining. The visitor may grasp at a glance or intuitively your point: signify by your attitude that you are impressed, that you find the person extremely intelligent.

You are not pouring information into a bucket. To the contrary, be on the lookout for contributions by your guest. It should be a dialog, not a monolog. You want input from that person. Refrain from a selling job, the most common mistake. Your visitor will offer suggestions and you ought to listen to them carefully. Who knows, s/he may be a reviewer of your grant application. Your most important task is to listen carefully to what s/he contributes to the discussion. Discussion is the key word here, there are two of you, a silent audience of one defeats the very purpose of the visit, namely the exchange of information and of viewpoints. Your visitor is a consultant, which you have the good fortune of having for free.

In other words, do not go through a standard, memorized spiel. This is the best way to give a poor impression of yourself as a salesperson rather than a scientist. What you hope for is to make a new friend. Aim at making this colleague your ambassador-at-large, singing your praise to the community. Equivalently, your purpose is to draw this person into your network of friends and associates.

Which brings up one final point, that of the parting gift. When the visitor leaves your office, do not load him or her with an armful of papers. Present no more than four reprints or preprints, the latter preferably. Your visitor will thus feel privileged, you have shared documentation ahead of publication, which is a symbol of trust and friendship.

REFERENCE: M. Alley (2003) *The craft of scientific Presentations.* Springer, New York.

a waste of time 

PROGRESS REPORT

Treat it neither as a waste of time nor as a formal obligation to a granting agency, or to the administration in your institution. Consider it instead as a golden opportunity to clarify your mind and to identify new goals for your research.

Let me give you a hint, which has served me well throughout my career: there is nothing an administrator likes better than to receive an unsolicited report. It means that you treat this person, not as a cog in a machine, but as an esteemed individual, whom you see fit to provide with information.

Start your report by summarizing in a short **paragraph** the substance of your previous report: what was achieved, and what your future goals are. Then proceed to state how you fulfilled the earlier objectives that you were able to meet. Do not gloss over the difficulties encountered. On the contrary, concisely describe how you solved them. In so doing steer clear of self-glorification.

In logical sequence, continue your report with earlier objectives that fell by the wayside, for whatever reason. List them briefly, indicating why you were unable to fulfill your program. If you came against an unforeseen problem, describe it. If you made a decision that a particular step was not worth the trouble, say so. If you realized that something else was more interesting, explain why.

The heart of your report is the third part. Discuss the results which could not have been foreseen in your earlier reports. This section should be at least half the length of your write-up. For each result, explain how it developed, describe its importance to the investigation, chart the new avenues it opened. Do not content yourself with raw results, also present at least an attempt at interpretation or modelization.

Which brings up the matter of numerical tables and **illustrations**. A research report is not a research paper. The former is meant for an administrator who lacks the expertise, even if a fellow scientist. Accordingly, present the data in predigested formats. This precludes inclusion of numerical tables. Conversely, be imaginative with illustrations. Make sure that the pictures or photographs speak to the non-initiate, if graphs are used make them simple, or, if possible, dramatic.

By the same token, only the skinniest of bibliographies is needed. Include references to papers which you and your team have published in this line of work since your earlier report. As for papers from your competitors, an extremely brief summary of the state of the field belongs in your introduction. It should be done in a few bold strokes, showing that you are aware of the evolving context for your work. It will also help to justify the redirecting that your research may have taken.

The conclusion to your text ought to first recapitulate the main points of the report, i.e., what has been achieved. Go on by setting the program for continuation. Try to do both in less than two pages. Having concluded the drafting of the report, number and entitle each of the sections. Write an **abstract**, which should consist in a maximum of half-a-dozen sentences.

Now comes the essential step, editing your report. Cross out anything unnecessary. Use a dictionary to help you find elegant wording. Allow yourself a single line in italics at the end of each section for a pithy and striking summary.

How will you know that your editing is a success? With a scale perhaps. Who cares for a weighty report of 35 pages? Try to keep the whole document below 20 pages, double-spaced.

REFERENCE: A.M. Penrose and S.B. Katz (2003) *Writing in the sciences: exploring conventions of scientific discourse.* 2nd edn, Allyn and Bacon Longman, Boston MA.

your best polished style

RECOMMENDATION LETTER

If you are asked or volunteer to write a letter in support of a colleague, who has applied for a new position, and who is eligible for an award or a prize, … the letter must be well crafted. Even more important, as a semi-public act, it reflects back on you. What you communicate shows your worth as a scientist and your ability as a stylist.

As with other recommendation letters, identify yourself in a **sentence** or two. Then proceed to describe how well you know the recommended person.

Now to the contents of the recommendation proper. Do not trust your memory. You need to refresh it with personal interaction with the colleague you are about to praise. If feasible, especially if this person has requested the letter from you, set-up an interview. It ought to last an hour or so. Ask questions, based on the resume which will accompany your letter. Write down the answers as the interview proceeds. Two questions which I have always found to be extremely productive are (i) what do you consider as your chief quality? (ii) Ask as the very last question, is there yet another question I have failed to ask of you?

A somewhat equivalent way to do it, especially if for some reason holding a personal interview is not feasible, is to ask the person to write the recommendation letter and submit it to you. Then, rewrite it entirely in your own words, in your own style, praising the aspects which the colleague glossed over because of modesty.

At this point, you have to sit and write the recommendation. It is necessary to go through at least a couple of drafts. Purge your text of all the stereotypes, of adjectives such as exceptional, creative, imaginative, energetic, innovative, …Yet, some adjectives can be crucial. To be an expert is trite. However, to be »a seasoned expert« is far more interesting to the reader and, if true, reflects favorably on your perceptions and judgment. Remember that understatement is considerably more effective than hype. Do not focus exclusively on the scientific merits. Address also the character of the person you are recommending.

Keep in mind that to a significant extent the modern recommendation letter is an American art. The reason is historical. The rise of the modern research university occurred in the United States in the nineteen-fifties and nineteen-sixties, at a time when American culture underwent one of its cyclical waves of hero-worship. This was the time when John F. Kennedy became president. Hence, your letter, while presenting in synthetic manner Dr. X's contributions to science, won't fail to express what you and others find admirable in them.

Do not exceed a couple of pages, anything longer is counterproductive. Writing in your best polished style is essential. Include at least a memorable formula, which will concisely communicate the merits of the individual. Be sure to save the file. Conversely, do not ever use any standard letter stored in your computer, it won't wash.

REFERENCE: L. C. Perelman, J. Paradis and E. Barrett (2001) *The Mayfield handbook of technical and scientific writing.* Mayfield / McGraw-Hill, New York NY.

unruly kids 

REFEREE REPORT

To review a paper for a journal, or a research proposal for an agency, are adult undertakings. Yet, most of us act irresponsibly. Too often, our behavior resembles that of unruly kids who enjoy hazing someone new to the block, who relish pushing the head of a colleague underwater. In so doing, true, they may trip up a competitor. However, the bigger damage is to themselves and to their reputation.

But should you answer the call to review a manuscript, in the first place? If you have a vested interest, pro or con, disqualify yourself.

Read the submitted manuscript carefully and annotate it with a pencil. Do it on the day of receipt: »do unto others etc«. I would recommend that you draft your report on the same day, that you let it sit for a couple of days prior to returning to the manuscript for additional scrutiny a couple of days later. You will be surprised yourself by how much this small delay will improve your text – not to mention the gain to your mastery in your field.

Before starting on your task, jot down your answers to a list of questions such as (most journals supply their own list):

- is this the appropriate journal for publication? If not, can I suggest a better medium?
- is the paper significant/important? Why? Why not?
- is it comprehensive with respect to its subject matter?
- are there omissions?
- are there mistakes? inaccuracies?
- is the work reproducible from the evidence provided?
- do some of the authors' assertions need to be qualified?
- does the paper conform to the same high standards as previous contributions from the same group?
- is the writing clear and fluid? Can it be improved? How?
- are there mispellings? typos?
- is the **bibliography** adequate?
- is the artwork necessary and compelling?
- is the **title** adequate?
- should the **abstract** be rewritten?
- can I suggest cuts in the manuscript?
- how can I sum-up in a sentence or two my overall assessment?

Organize your report. Order and number your remarks in sequence. Quote page and verse for each of your comments.

Since the rule of the game entails negative criticism, you have to abide by it. However, your remarks will weigh all the more heavily when you provide also positive comments. The colleague whom you may need to excoriate (if need be) for having submitted such a flimsy piece of work, such a shoddy rag, will heed your advice all the more readily if your review also has a flattering bit. Make sure to include at least one sentence, better a paragraph, recognizing the expertise, paying homage to earlier contributions to the field from the same author(s). This is self-serving: the editors will take note that you are acting as a responsible reviewer. Somehow it will get into the record, in your credit account.

How long should your text be? A report to the editor of a journal should stay within a couple of pages. A review of a research proposal, one of a book manuscript for a publisher may take half-a-dozen pages. Those are only guidelines. In any case, crisp and clear writing is important. Even more so is a measure of wit, for buoyancy to your text and making happy your initial reader, an editor.

Last but not least, make your report timely. If there is any doubt in your mind that you will answer promptly the assignment, decline the honor. To be asked to review a manuscript and to answer by being inadvertently tardy, worse by dragging your feet deliberately in order to slow down publication by a competitor, are unacceptable. Such behavior is certain to ruin your reputation to the editor. Somehow, the word will go out. You will be known as a bad guy. You'd better claim being too busy to give proper attention to that manuscript, and suggest someone else as a referee. Cooperation is valued, obstruction will be obvious and it will only give you a bad name.

REFERENCES: D. Hamermesh (1992) *The young economists's guide to professional etiquette.* Journal of Economic Perspectives 6(1):169–179; (1992) *Facts and myths about refereeing.* Journal of Economic Perspectives 8(1):153–163.

Demonstrate mastery

RESEARCH PROPOSAL

This is the all-too-familiar system by which people trade ideas for money. Actually, the terms of the exchange is funding in return for the promise of results. Clearly, it is similar to the stock market. The writing of a proposal thus resembles that of a bond issue.

Originally foreign to science, it came in the aftermath of World War II, with the rise of research universities. Former military personnel turned to administration of science, either corporate or academic. Spontaneously they used the format of military missions, the spelling-out of objectives and means. Mission-oriented research was thus developed. Whether good or bad, it has remained with us. It has sadly led to »me-too science«, imitative and conformist.

There are tips to make your application be successful. There are specialized commercial outfits which may help you to put together your grant application, but they cannot outperform the combination of clear, logical thinking and an effective style. A successful research proposal needs an important idea to seed it. This is a necessary, not a sufficient condition.

The way in which you express your concept will either sidetrack your application for a grant, or it will lead it along the avenue with leads from an uppermost classification to funding.

The main mistake scientists make in writing grant applications is to imitate their standard publication mode, replacing the present tense with the conditional. This trait is all the more tempting, except that for the experimental part, the proposal is structured in much the same way as a publication. An **introduction**, a **conclusion**, a description of apparatus and procedures, a discussion, and a **bibliography**, the latter preferably comprehensive.

Instead of such a poorly presented approach, start with a summary of prior work in the area, paying due respect to your predecessors and your rivals. State your prior contributions in a modest manner, steering clear of hype and one-upmanship. State the problems you want to solve and the original approach for doing so. Demonstrate mastery without claiming it.

In writing a grant application, most resort to the common ploy of presenting desirable results already obtained. Be aware that your reviewers will take it for granted that you already have done the first phase of the proposed research. They are not fools. You might as well be truthful. Describe your exploratory work and the inferences from it. State the leads these initial observations suggest, the complementary experiments needed as extensions and for controls, the hypotheses to be tested. Remain concise, try to stay within a maximum of twenty pages, exclusive of the references, proposed budget and questionnaires.

REFERENCE: G. E. Kennedy (2001) *Professional and technical writing: problem-solving at work.* Prentice-Hall, Englewood Cliffs, NJ.

a recipe for disaster 

RESEARCH TALK

The research talk is a genre declined in various ways, an oral presentation at a conference, an invited **lecture**, a keynote address, and so on. Adjust to your given audience. Never use an unmodified standard presentation, you would sound like a tape recorder. It is a matter of courtesy and of efficiency. Attention works both ways. By being aware of the needs of your public, you have a better chance of addressing them and of, thus, being not only listened to, but heard.

First, find out the length allotted to your presentation. As a rule of thumb, both the one-hour lecture and the 10-minute bit are relatively easy to deliver effectively, even brilliantly. The most difficult talks, at least in my experience, are the 30-minute ones. Compressing a lecture you are used to presenting in an hour to half that duration is a recipe for disaster. It requires intensive preparation to deal with such a time slot.

As you are about to start after having been introduced, make sure to have a nearby glass of water and throat lozenges in case you dry up, a pointer – a stick rather than a laser, which may emphasize trembling in your hand – and a handkerchief to wipe your brow or blow your nose.

Test whether people in the back row can hear you. It guarantees that you will be audible to everyone. Speak slowly, much more slowly than your usual pace. Everyone will be grateful for the careful delivery. They will be able to follow you more easily and to take notes, a behavior to be encouraged.

Start by telling people what you are going to say, then say it, and then devote a couple of minutes at the end to recapitulating what you have said.

Audiovisuals are great but do not become their slave. Remain capable of a blackboard-and-chalk presentation. Make sure that your slides, transparencies, or videos are visible without having to dim lights – which is conducive to people falling asleep.

Avoid showing tables of numerals. Graphs present the trends much more effectively and more pleasurably to the eye. Insert some amusing or beautiful slides, the audience will be grateful for such breaks.

Never attempt presenting ALL your results. A sampling is sufficient. Have at the ready all of the data to answer questions posed during the discussion.

Never read your text. This is also an inducement to sleep.

Do not stay glued to the lectern. Move about. Be a little physical. Gesture to emphasize a point. Be dignified, in all ways. You can dress informally, but not too much so. Have a modest demeanor. Be communicative in a friendly way. Avoid histrionics. Do not overdo the **body language**.

Being well-prepared is the key to success. Rehearse beforehand. Time yourself. Learn your opening and closing **sentences** by heart. This will allow you some precious spontaneity.

REFERENCE: J. v. Emden (2001) *Effective communication for science and technology.* Palgrave Macmillan, London.

a mountainous landscape 

REVIEW ARTICLE

How to write a review article or chapter, without the assignment turning into too big a chore for you and into a tedious task for your readers? That is the question.

You will have to do a very thorough job. A review has to be rather comprehensive. Some review journals are demanding and you will need to cite every single piece of work in the field. Other journals are more lenient and will allow you to focus on your own work. Your colleagues will be unhappy if they don't find their name cited, whatever the reason. You want to avoid causing any such slight omission. Furthermore, your review article needs to remain totally objective.

Let us query the meaning of this adjective before we go on. Because we are all, or nearly all, extremely vain, we equate objectivity and an equal treatment. Such an egalitarian policy may be diplomatic. However, it will doom your paper to a library catalog look alike, a stack of file cards.

A good review article, conversely, will single out – let's say 20 – major contributions to the field which have appeared during the period under review, typically the last three to eight years. For each such important paper or series of papers, explain why you deem it significant, why you consider it as a breakthrough and describe or allude to its impact.

Let us assume then that you have taken care of the above. Your review is now akin to a mountainous landscape, with a few peaks standing out. Are you certain of having been truly objective in your assessments? Could you have been blind to some distinctive achievements which have somehow failed to come to the forefront, to stand out in the public eye? Your role as a reviewer is also to give them the prominence they deserve. I would even argue that such resurrections are your most important task as a reviewer. They are likely to become the best asset for your paper.

You may now find yourself at the end of a first draft, weighing perhaps 35 double-spaced manuscript pages and bearing 250 references. The vast majority have deserved the one- or two-sentence treatments, while devoting a whole **paragraph** to those papers singled out for their importance.

It remains for you to polish the write-up. Outstanding writing is the only distinguishing feature which will make your manuscript shine. But think of the rewards for your hard work. It sets you in the potential posture of a field leader. For instance, **research proposals** will be referred to you, a mixed blessing. Hence, you will benefit from good sources of information (the horse's mouth, so to speak). You will have your finger on the pulse of the field for at least a couple of years. A big reward is that you have organized the field in your own mind and probably have a very good idea of what needs to be done, so you can work on it in your laboratory. A more distant reward is for future historians to rely on your review to identify lines of force in your particular field.

In so doing, you will have written an authoritative review and stepped on a few toes: are there greater satisfactions?

REFERENCE: P. Rubens ed. (2001) *Science and technical writing.* Routledge / Taylor & Francis, UK.

beyond such a dissection 

SEMINAR

Presenting a seminar or attending one are different roles. To contrast them as active or passive is mistaken. Both are equally active if the seminar works as it should.

When presenting a seminar, speak slowly. Be deliberate, remain very much aware of the impact of your words on the audience. Identify your assumptions clearly, from the outset. Do not bury your audience under data. You do not have to show all your results. Prepare a careful selection, and offer your interpretation, having separated the raw data from the model(s) used in its analysis. Explain the implications you see from your results. It would be foolish on your part to expect to be greeted as the Saviour. All the parts of your talk may be challenged rather than being accepted eagerly. Your work is there, laid bare and about to be picked apart, criticized relentlessly. Remember, the purpose of the seminar, beyond such a dissection, is to legitimize and validate your work through its public discussion. The seminar attendees are not out to get you, they have nothing against you personally. Listen carefully to what they tell you, take note of it. You will find that it helps you to improve your work markedly.

If on the receiving end, your first task is also to take note. Write down the information carefully. Be selective. Focus on the parts which you feel may come up in the discussion. Of course, also record what is likely to be important to you personally at some point in the future. Your notebook may then be your only access to such documentation (under the label of »personal communication«).

A seminar has the same goal as a day in court, namely to establish the truth. The parallel to legal proceedings, even though it clearly cannot be pushed too far, is nevertheless instructive. The seminar is an adversarial procedure. It is likewise conducted as a spoken discussion. The presenter holds and defends a version of the facts. The colleagues who behave as the devil's *advocate* and attack it, present their own selection from the data, and their own interpretation.

The word »seminar«, a cognate of both »seminal« and »seminary«, means etymologically a seeding. The attendees are seeding doubt, and doubt is the touchstone for truth. In doing so, the protagonists not only get under your skin, or flay you alive, they are testing the truth-value of what was presented. The seminar is a method for getting at the truth through dialog, first used, primarily as a teaching device at the University of Berlin founded by Alexander and Wilhelm von Humboldt in the first decades of the nineteenth century.

When a seminar works, the most important part thus is the discussion. How should it be run? You may be called upon to do it as a moderator – a self-defining term. Alternatively, you may have to do it from the podium, as the presenter. Questions may have to be repeated louder. They may have to be rephrased more articulately. Yield the floor to people asking questions or offering comments in the very sequence in which they have manifested themselves (important). You may have to raise the first question, in order to get the discussion going.

It is necessary for all the viewpoints to be recognized and to be expressed. A thoughtful moderator may have prepared in advance, based on the topic of the seminar, an agenda of questions to be raised. He or she may offer a framework for organizing the discussion. If in the course of the discussion anyone obstructs the flow of ideas – not addressing the issue, going off on a tangent on a pet topic, being unnecessarily aggressive – , it is the role of the moderator to intervene, and to turn to someone else. A discussion of a one-hour presentation ought to last at least ten to fifteen minutes. If the topic warrants it, prolong the discussion for a commensurate time, one hour too. If you are one of the discussants, each of your interventions needs preparing. Write down in your notebook, or on the palm of your hand, the various points you wish to make. Stick to simple language. Be terse. Remain courteous to the seminar speaker. Listen carefully to the answer. If your question is not being addressed, say so politely. Like everyone else in the room, you are seeking out the truth of the matter.

A seminar has a single duty, to try and give a truthful account of reality. It has considerable educational value. Everyone learns from a good seminar. Its role is to seed unexpected ideas.

Likewise, a seminar runs an unpredictable course. The intellectual flow is both above ground and underground – an idea may seep in slowly and enrich your own work, even though at a distance from the ostensible topic of the seminar.

Be not shy. Controversy is good. It is the lifeblood of science. For the seminar presenter, to run the gauntlet of questions and comments may feel like an ordeal. Or a bullfight, with the risk of being gored and of feeling one's ego mutilated in the aftermath. But it is worth it.

The argument is usually about interpreting the available facts, not about their validity. Coexistence of diverse viewpoints is healthy. It goes against dogmaticism. The seminar remains an institution central to science and to its communication.

REFERENCE: J. Davies (2001) *Communication skills.* Pearson Higher Education, UK.

attractive to the eye 

SLIDES AND TRANSPARENCIES

This particular medium is ancient. Slides existed already in the eighteenth and nineteenth century, made out of painted glass and projected with a magic lantern. With the invention of photography, slides became ubiquitous. In mid-twentieth century, there were two main kinds, 35mm slides made with ordinary photographic cameras, shown with the ever-present Kodak carousel projector during the whole period 1960–1980, and Polaroid slides.

An important part of the process, usually performed by a professional photographer, entailed production of a positive print from the original negative. I remember also shooting 35mm film, and coloring by hand the lettering on the back of the negatives. You could also use recipes for development, so that you would go from a black-and-white original to, say, white lettering on a blue background.

Transparencies made use of another piece of equipment, also dating back to the second half of the nineteenth century, the overhead projector. Rumor has it that transparencies were devised by physicists at the beginning of the nineteen-seventies, during their conferences. One would grab a piece of transparent plastic, vinyl or acetate, and scribble an equation on it with a marker pen. The transparency was sister to back-of-the-envelope calculations.

From such informal beginnings, the medium then gained in sophistication, when personal computers became widespread, and one could print in several colors. It took about a decade for such electronic output to become routinely broadcast directly from the computer (PowerPoint became the universal software for such use in the 1990s) onto the projection screen. With such technology, one achieves nowadays results comparable to those from an in-house graphic artist and a photographer 20 or 30 years ago. The technology has made graphic artists redundant (an awful term) for this kind of work.

Arguably, the originators of the slide show were art historians. I have always envied them. They deal with admirable material. Not only are the artworks they display attractive to the eye, the lecturer interprets them with you, explaining the symbolism, walking you through the piece to make you understand the intentions of the artist, explicit or more hidden. Their slides talk to the eye and to the mind both, an ideal other fields would do well to strive and emulate.

The other origin of presentations using slides and transparencies, as both quicker than writing on a blackboard and allowing one advanced canning of the material, is in the courtroom. Expert witnesses, attorneys no doubt following the example of those scientists called upon to testify, would use these media to present their evidence to the judge(s).

Such visuals as the slide and the transparency allow you to put up the evidence for everyone to see. Meanwhile, they listen to your spoken commentary, i.e., to your interpretation of the data. You may be showing documents, diagrams, models, statistical evaluations, pictures of the laboratory and of your coworkers, and so on. These are alike so many legal exhibits.

Your goals, in going through a series of slides or transparencies, are (i) to tell a narrative, and (ii) to convince the audience of the validity of your approach. I have already mentioned the assets of this medium: it is quick, inexpensive, printable, and above all, self-made. Were you called upon to give a lecture on your work at 5 in the afternoon and need 25 slides to illustrate it, you might be able to start preparing them at 9 in the morning and to be done by noon.

The medium repays with mediocrity such ease in preparation. A talk based on a PowerPoint presentation is all too likely to be shoddy: visuals crammed with too much information, shown too quickly for a listener to note the most interesting parts, and accompanied with an uninspiring because poorly prepared talk.

Remember that your audience can read. People will read the textual parts of your slides quicker than you can read them aloud. Hence, you should provide a spoken counterpoint to what is written on the visuals: complementary, qualifying it, extrapolating from it …

Your audience can read, true. But are the people attending your presentation able to actually read your slides? Try to avoid mere photocopy, the scrawl, sanserif fonts (fonts with serifs, i.e., the tiny pedestals at the foot of the letters, are easier on the eye), too many capital letters (they are also more difficult to read), italics (likewise), unfamiliar acronyms, a surfeit of information, …

Keep in mind that your presentation is an uphill struggle. It runs against indifference, information saturation, skepticism, disbelief, the all-too-frequent poor acoustics (as in hotel ballrooms turned into lecture halls), and the lurking sleepiness of people held captive in the dark. Accordingly, never use a dark background for your visuals, make them easily viewable in a non-darkened room, contributing to the general lighting.

You need to appeal to the so-called Broca area in the brain of your listeners. Jolt your audience. An unexpected picture will rekindle interest. Breaking the rhythm, itself easily narcotic, for the telling of a relevant anecdote may help to relieve the boredom. You do not have to use up the whole time allotted to you. Allow an extra six minutes: as you prepare to start, the session will already be running late; the chairman will take a couple of minutes to introduce you; you may be stopped in the midst of your presentation by a question or by some incident you will need to react to; why not read or better quote a poem? Why not sing a song? Your lecture will thus thrive on the unexpected, especially if it sounds spontaneous.

The telling of a joke, this trick of sales people and of politicians during the first few minutes of their talk in order to pull people to their side, in order to penetrate through the normal mental defenses of forced listeners, is a deeply ingrained American habit. It suffers from being too mechanical. However, telling jokes is not only acceptable, it is to be recommended – provided that the joke has something to do with the rest of your narrative and spices it in an alluringly funny way.

REFERENCE: E.R. Tufte (2003) *The cognitive style of PowerPoint.* Graphics Press, Cheshire CT.

Superficiality is deadly 

STATE-OF-THE-ART REVIEW

What are the differences between a **review article** and a state-of-the-art review? The main one is the viewpoint. The former demands only competence, the latter requires vision – vision ahead foremost, and a sense of history.

Who will read your piece? There are two main but totally separate constituencies. The first is made of senior scientists, leaders in the field covered and who are likely to be critical. The second is made of young scientists entering the field and turning to your review article as an educational guide. Catering to only one of those groups and failing to consider the needs of the other is an extremely common mistake.

You may forget about general readers entirely. Only a small fraction of your readership, less than one in five, will read your review as a matter of both general interest and curiosity.

Why write such a paper? You will have been requested to do it by the editor of a journal. Clearly, you would be performing a public service in writing a superior review. Not only will you be helping others to gain an overview of the field, the position of eminence this implies will project back on you, most favorably.

How to write it? The greatest obstacle, in my opinion, is that mastery of a field and objectivity in reporting do not necessarily go hand-in-hand. The former gives you a largely illusory eagle eye's view, you have long identified in your mind the leaders and the plodders. Accordingly, your review willingly-nillingly will discriminate among them. As a critical reviewer you will necessarily describe developments you deem important in more detail than those that are not, thus automatically adjudging merit. Indeed this is one of your functions. But leave any kind of a final judgment to historians, you are too close to the current action for a bird's eye view.

Hence, you need to write an objective review, one which will give its due to the leading players in the field, as well as to the minor players, who bring in unexpectedly those crucial fringe or marginal benefits. Remember, scientific discoveries seldom happen in the mainstream. You are called upon to provide a combination photograph and radiograph of the mainstream. However, somehow you need to grab as well all of the incidental results which, who knows, might realign the whole field in another direction?

Of traps, there are quite a few. Superficiality is deadly. If you are tempted, out of laziness, to compose a mere annotated bibliography or to compile a file index in print, those just don't qualify. One-upmanship is another all-too-common fault.

The three foremost assets of your review will be comprehensiveness, objectivity and authority. Only by penning an authoritative summary of what the field currently resembles, of where it is likely to be headed and of where it ought to be headed will you be fulfilling your responsibility.

REFERENCE: C. Harkins and D.L. Plung (1982) *A guide for writing better technical papers*. Wiley, Hoboken NJ.

well-prepared protagonists 

TELECONFERENCE

The scientific community is dispersed all over the world. One of us may have more in common with another scientist at the antipodes than with the colleague down the hall. Accordingly, some collaborations are helped by sophisticated communication technologies.

Teleconferences are one such means. They are conversations among several people, in real time, which dispense with the usual requirement of participants being in the same room. They can be anywhere.

This form of communication can thus save considerable time, traveling and money. It may use channels such as the telephone, the broadcasting of the face image and voice via telephone lines and/or satellite transmission, the Internet. In this last case, the format is the chatroom supplemented with images from webcams, i.e., small cameras perched on top of individual personal computers.

Teleconferencing can be more or less successful. There are technical difficulties. There may occur an annoying lag time between the two transmissions of image and sound. Each of these two channels privileges speed to quality. Voices may be a little distorted. The poor definition of the image may pose a problem. In any discussion, nonverbal cues are essential. Due to the poor quality of the image, such **body language** may elude perception. Just think of the personnel present in a television studio: cameraman, sound engineer and other members of the technical staff. A standard teleconference is unable to draw on such professional expertise. It shows!

There are human problems, in addition. A single participant can torpedo the proceedings, by being out of tune: inattentive, having people repeat what everyone else has already assimilated, interrupting, speaking out of order or not to the issue being discussed, …

The best teleconferences gather well-prepared protagonists. They occur within a closely knit group: people who already know one another well, are used to working together as a team, who are familiar with one another's idiosyncrasies.

It is best to be dealing with a small group. The effectiveness decreases exponentially with more members. Give careful consideration as to whom you ought to include.

A coordinator is needed, somone who will direct unobtrusively the flow of the conversation. This chairperson will prevent simultaneous interventions from mutually interfering. At the same time, this person will put on the air anyone having something to contribute. She will know when to drop stage management in favor of spontaneity.

In the best cases, the teleconference gathers people who are apt to operate on the same, common level. It demands an egalitarian rather than an authoritarian or hierarchical mode of interaction.

REFERENCES: G. F. Hayhoe (1993) *Communicating across the country or across town: planning and producing effective teleconference meetings.* Technical Communication 40:160–163; K. Kelleher and T. B. Cross (1985) *Teleconferencing, linking people together electronically.* Prentice-Hall, Englewood Cliffs NJ.

Include some whimsey 

VISUALS (FOR A LECTURE)

To seduce through the ears and to appeal to the eyes, those are your tools for a talk using visual aids. What are those? Typically slides or transparencies, they alloy images and text. Sometimes, they consist only of images, or just text. Aim at about a slide per minute, about 60 for a one-hour talk; for transparencies every two or three minutes, which translates to about 25 for a one-hour lecture.

Include in your presentation the recurring indication of the advancement of your talk. It is important that your listeners be given such information: how many more parts to the talk, how far from the summing up, when is it going to be over? Show a table of contents at the start, return periodically to it to indicate what remains to be covered.

My first rule is the most difficult to implement: make your **visuals** and your speech complementary. Their occasional redundancy is good. But avoid duplication becoming systematic. For instance, avoid reading aloud what is written on your slides. Your audience can do it much more quickly silently. Paraphrase the words on the screen. This is easy. However, to paraphrase creatively, in a witty and seemingly spontaneous way, is quite another ballgame. How to achieve it? By practice.

The second rule is substraction. Once you have assembled a first draft of a presentation – using PowerPoint for instance – delete at least 20% of the material from individual slides/transparencies. You will find them quite a bit more effective.

For the text, select a single font for readability. However sexy they may look, sanserif fonts such as Arial or Geneva are hard to read. The serifs are the small slippers which letters bear on their feet. Their implied horizontal continuity helps the scanning eye. Personally, I favor a serif font such as Palatino or Times.

A fourth rule is to strive for continuity, built-in unobtrusively. Be attentive to the dominant color in your sequence of visuals. For instance, you might start with a light pink background and, rather than sticking to a uniform background for the entire presentation (often advisable) gradually work your way to a deep red. Speaking of such color cues, be attentive also to their psychological, i.e., cultural connotations.

My fifth rule is to maximize the iconic versus the textual content. Tables of numbers are an absolute no-no: only display your data with plots. Remember to show just a fraction of your data (Rule Two).

And what should your images be? Obviously, they need visual interest. Show images, which are a compromise between harmony and a jolt; the aesthetically pleasing and the informative; the obvious and the enigmatic. The mind enjoys having to do a little work – a little, not too much – when reading the picture. This is good. To gain understanding actively rather than as a passive viewer gives pleasure.

Include some whimsey. Don't hesitate to be a little outrageous. Back in the 1960s, before permissivity became widespread, during a conference of advertisers one of the speakers introduced a few mineralogical pictures among his slides. They just flashed through, with no comments on his part. His talk was a major success. In other words, keep your audience alert by, occasionally, showing it the unexpected. As in the above anecdote, the occasional surprise material, shown without comment, will spice up your talk.

Not only fantasy, beauty is also an essential ingredient. Ideally, each talk ought to include one piece of art capable of being exhibited separately, because it can stand on its own.

What about the twin projector mode, setting-up a counterpoint between two images projected together, next to one another and occasionally overlapping? Such an ambitious undertaking is worth the additional effort – and money, if you have the show prepared by professionals, which I recommend – only if the actual product shown to your audience is perfect. Moreover, it makes for a more passive audience: are you sure you really want it?

REFERENCE: J. Krasner (2004) *Motion graphic design and fine art animation.* Principles and Practice, Elsevier, Amsterdam.

an alert text 

WRITING A BOOK REVIEW

Prospective readers of a **book** depend on this genre, within science communication, for their decision to buy the book. If you are called upon to write a review, the best way to be worthy of the honor is to be very conscientious in your reviewing. What you write will answer questions for the reader's sake, questions such as:

- is it a good and easy read? Is the writing pleasant, engaging?
- is it likely to be and to remain the reference work on this topic?
- is the author knowledgeable about what s/he is writing about?
- are there mistakes? (in the affirmative, identify and correct them)
- are there omissions? Are they truly detrimental to the book value?
- what kind of a need does this book fulfill?
- what are the other existing books on the same topic?
- is it worth the money? Is the quality-price ratio decent?

In the early parts of the review, summarize the contents of the book. Then, answer all the above questions. Comment on whether the author has done a good, an *authoritative* job with lasting value.

You may also want to outline, since you are presumably an expert too, how you would have gone about treating the identical subject. But only after having given the author the courtesy of first analysing his work. This is where your review will start to take off. This is where it will lift from a dry and boring library card.

A book review is a window on your expertise. It is the place to wear your scholarship lightly, with elegance and wit. Tell an attractive anecdote likely to draw readers and buyers to the book. If your review is negative, you may nevertheless want to make the topic of the book attractive, since it is your specialty. Make your writing lively and entertaining.

A book review should be an alert text, one to make the reader feel excited and sharper.

REFERENCE: H. E. Neal (1964) *Nonfiction: from idea to published book.* Wilfred Funk, New York.

PART II

THE GENERAL PUBLIC

GUIDELINES

ANECDOTAL

»One day in 1665 in the small Swiss town of Appenzell, Ignatius Brülisauer let off a shot from his pistol directly in front of Hans Joseph Wälti's house. The very pregnant Frau Wälti got the brunt of the shot directly in her face as she stood watching at the window with a small child in her arms. The Paracelsian surgeon and village doctor, Umrich Ruosch (1628–96), removed the shot and charged the pretty sum of fifteen doubloons. Ignatius Brülisauer took Ruosch to court for overcharging and the court called in three doctors of medicine and two barbers from the surrounding region to judge the true cost. These experts determined that the wound could not have been severe since Frau Wälti had neither dropped the child she was holding at the time nor suffered any complications when she gave birth not long after. They estimated that the treatment of such a wound would be more in the range of one doubloon. To this they added the costs of messengers and doctor's visits, but they could not say how much the medicines Ruosch had prepared would have cost because, despite their repeated queries, Ruosch would not let out the secret of his preparations. In the end, the court set the fee for treatment at ten doubloons and divided court costs evenly between the two antagonists in order to dissuade them from bringing such matters to court again. It did not actually prevent Ruosch from returning to court. Indeed he appears to have been a self-confident individual not shy of confrontation.«

I have reproduced the entire first paragraph from a book review (a modern edition of the alchemical handbook which Ulrich Ruosch carried with him) by Pamela H. Smith from Pomona College, in Claremont, California, which appeared in the *British Journal for the History of Science.* 37(3):348–349 in 2004.

The anecdote she relates is a lovely entrance to her text. It seduces us, the readers, because it recaptures long-gone moments. It has the flavor of daily life. From the opening **sentence**, we are under the spell because it sounds like the start of a bedtime story: »Once upon a time, there lived in the town T a person by the name P etc.« In other words, the anecdote sets the backdrop to the handbook which Dr. Ruosch had put together for himself, which is the topic of the review. It establishes the historical context: small, German-speaking Swiss city, seventeenth-century, barbers serve as para-medics, litigation is endemic, the money unit is the doubloon, and, most important for what is to follow, a physician is able to defend the drugs of his own preparation as proprietary knowledge.

There are many kinds of anecdotes, the historical shown here is only one kind. The virtue of the anecdote is to bring life into a text. Its main drawback, some people feel, is that it detracts from the more serious issues. But such a reservation is easily countered in practice. It hinges on the transition from the anecdotal to the general. The author has to take some pains to establish the anecdote as representative. In other words, the main virtue of an anecdote is to offer a concrete example. Any discussion is improved by being thus tied to the real.

The word »anecdote« goes back to the Byzantine historian Procopius (ca. 499–565 AD), who published under the title *Anekdota* gossipy memoirs of his day. It was resurrected by Voltaire in the eighteenth century. Sure, anecdotal stuff shouln't be taken too seriously. But this does not mean one ought to refrain from using it, nor from occasionally drinking champagne: both make us feel happy and light-headed.

REFERENCE: J. Kirkman (1992) *Good Style: Writing for science and technology.* E. & F. N. Spon, Routledge UK.

our triumphs 

AVOID TRANSLATION

The most common misconception of science popularization is to treat it as mere translation, from the technical language of a discipline to the common language. Do not misconstrue me: of course it is absolutely necessary to forgo recondite terms which only specialists can understand. What I'll be arguing here is that this is far from being enough.

Science journalists occupy that gap. They see themselves as the natural go-betweens. They know how to put the results from scientists into simple terms, understandable by anyone. Are we scientists not capable of making that effort? Again, I shall argue that this is far from being enough. Before going on to what is needed, let me emphasize why scientists are superior to reporters, why there is no real need for intermediates, for translators between scientists and the community at large. Scientists know what they are talking about. They know it first-hand. As a rule, hands-on knowledge is preferable to hearsay.

The reason why a translation is not enough is clear. Just look at a scientific publication, any science paper written for one's peers. It does not consist of terminology only, a set of complicated words sounding totally foreign to any outsider. It also carries a syntax. Even more important, it holds a **rhetoric** for convincing colleagues that (i) the work was performed according to the rules; (ii) the results are believable; (iii) the conclusions are important and they open-up new avenues of research. To this effect, we present our peers with evidence, as if in a courtroom.

None of that is readily translatable. Actually, most often it is impossible to translate. And we know darn well that omitting syntactic and rhetorical aspects in communicating science results to the public, as the science journalists do, is self-defeating. Our triumphs, small and big, are flattened. Their juice oozes out.

To counter such a grievous loss, science journalists play on base instincts. People can be selfish and self-absorbed, even morbidly so. Why should the man in the street be interested in a piece of scientific work? For utilitarian reasons, if he or she has the feeling it will affect his or her life and well-being. Hence, journalists devise and stress artificial connections to materialism, to one's health and prosperity.

What is needed is not a translation. It is a re-creation and should be recreational as well. The order of the day is not to translate from Martian into English. Rather, one needs to rewrite the whole message into English. But not into any English: the need is for a prose totally different from our usual mode of expression. Whereas the scientific prose, that of the papers we write for journals, is objective, impersonal and argumentative, prose for the layperson must conversely appeal to emotions, bringing in the human factor. It should be inviting. Done well, it will summon the reader inside the laboratory, to watch us at work.

We scientists are used at making assertions preceded by all the conditions, restrictions, etc. to validate the statement. It makes for a leaden prose. The lay person considers this habit unnecessary and tedious. It is part of our training to write for example, »in New York City, in the midtown section of Manhattan, it was raining at 4 PM on October 15, 2004«. Most readers would content themselves with the admittedly less accurate »it was raining in New York«.

However, necessary as it is to keep the information terse and to the point, it is still not sufficient. To limit information to the results only, as science journalists are wont to do, without telling the why and the how and the where we go from here, lacks the true essentials. It throws the baby out together with the bath water.

But to attempt to translate, rather than to re-create the procedural aspects is unwieldy. Often it is unfeasible. Sure, what is called for is hard work. You will have to start from scratch and to write a brand new publication, in a format highly unusual for you. But it is well worth the effort.

REFERENCE: T. A. Rees Cheney (1991) *Writing creative nonfiction: fiction techniques for crafting great nonfiction.* Ten Speed Press, Berkeley, CA.

welcomes such interruptions 

BODY LANGUAGE

Turn it into your chief asset. Make it switch your audience from feeling suspicious and worried to becoming engaged and trusting. You will hold them in the cusp of your hand.

Your purpose is to distance yourself from the stereotypes about scientists, as dull, dead-serious, aloof, distant and even arrogant, decidedly not fun fellows. Accordingly, to slaughter such prejudice, show your openness, your modesty and simplicity, your cheerfulness and enthusiasm. Your body will signify those qualities.

For starters, dress like most of your listeners. When in doubt, go for a more casual attire.

Be at your most pleasant. Feel relaxed. It will help everyone. Avoid using the shelter of a podium and/or lectern, unless you have no other option. At the outset of your presentation, scan your audience and make eye contact with some of them. Keep doing so during your performance. Smile, often: not a forced smile pasted on your face, but a genuine one. Isn't it a pleasure telling people what you enjoy doing most?

Make occasional fun of yourself, but with a light touch, avoiding self-denigration. From time to time, open your arms in a wide, embracing and welcoming gesture. This will also help your delivery by facilitating deep breathing, an obvious essential to public speech. Speak clearly and speak slowly.

Slow down. Give yourself the impression of having cut your usual pace by a factor two. The actual reduction, likely to be about 30%, will make your listeners more comfortable.

Don't be rigid and stilted. Physical exercise earlier in the day will greatly help in making you more supple, more nimble too, for the occasion. Be expressive, with your face and your hands especially. In the perception of your public, those two features alone amount to 80% of your body.

Be aware of the impression you give. Instantly correct what may go wrong: too loud? not loud enough? too immobile or, conversely, gesticulating? a monotonous delivery? main points not well emphasized?

In anticipation of the discussion at the end, turn yourself into a good listener too. Someone who does not mind being stopped at any time during the presentation, a feature which you will have announced at the very beginning. Someone who welcomes such interruptions, and expresses visible delight in being asked questions. Remain aware that the persons who have come to listen to you are interested in YOU even more than in what you will be telling them.

REFERENCE: S. S. Feldman (1969) *Mannerisms of speech and gestures in everyday life.* International Universities Press, New York.

a twist to the story 

BUILDING A STORY

Building a story is necessary since efficient science communication cannot be a mere translation from the language of science into the vernacular. A full reconstruction is needed. A text is almost always needed, whatever the final medium. What the public expects is a narrative. It wants a story that appeals to the subjective and the emotions, rather than one remaining squarely within the dispassionate cerebral and objective world of scientific ideas.

Quite a challenge, isn't it? Try starting with the **illustrations**. Once selected and put into sequence, they ought to tell their own tale. After you have established this iconic thread, and only then, you can start writing the accompanying text as a separate version of the same story.

For that purpose, start by noting the points you want to make. Order and number them in logical or in chronological sequence. These will be the parts of your story, their roots. The chronological order is often the easier, the more natural to follow. Define a locale: your story has to be set somewhere. The laboratory is an obvious choice, there are others. Choose and define the characters in your story, who does or did what. The characters may include inanimate objects of your investigation.

Do not take central stage yourself. Remember: you are the master puppeteer, s/he who pulls the strings, and not the most important among the puppets.

Think hard about whom you are addressing. Which segment of the public are you targeting? What are its needs? How best can you answer them? Can you involve, not only things, but also people in your story? Human interest will boost the value of your tale.

Who were »the bad guys?« In other words, what came between yourself and the coveted new knowledge? Such resistance to understanding might become the core of your story, once converted into episodes, within the flow of the successive events you are narrating.

Among such events, was there anything unexpected or surprising? Tell us about it. Was there a twist to the story, some revelation big or small that you might save for your ending? If indeed, as is likely, things took an unexpected turn at some point, what exactly happened? Tell it to your readers.

Is there a take-home lesson? Spell it out. How do you know this is the end of the story rather than it being just one in a series?

Why did you choose to tell this particular story? Do you have a good **title** for it?

Can you see it turned into a **TV show**? If you answer in the negative, you ought to immediately go back to the drawing board.

Was there an enigma, a riddle to be deciphered? How did you accomplish it, what gave you the clue or the key? This will be a crucial feature in your reconstruction for the general public.

Finally, has your story some built-in amusement, such as funny situations? Why not? Find some.

REFERENCE: P. Rubie (2003) *Telling the story: how to write and sell narrative nonfiction.* Collins, New York.

the forward look 

CONCLUSION

Coming to an end simply and elegantly is the last image of yourself you will leave your public with. Hence it is worth chiselling into an attractive sight. How then should you take leave of your readers? There are a few options, which I shall now outline in order to provide you with a choice.

First in line is the easy way out: content yourself at the end of the last paragraph of your text by capping it off with one or two brief sentences, and these should preferably be generalizations.

Science as problem-solving: your text may well have taken such a tack. You have recounted and it will make for fascinating reading, how you were able to find the solution (or not). If the question remains open, unsolved, just say so, it will make an adequate conclusion to your story. Conversely, why don't you raise at the end a question yet to be solved, one similar to the problem-solving you have just narrated.

A widely-used device for a conclusion is the prospective. You project yourself into the future and you prophesy. You are announcing what is likely to be the »future work in this area«. You announce the kind of conceptual breakthrough one might expect: »it may be possible to learn so and so«. You spell out what some of the applications will be, with a catch-all phrase such as »we expect it to see wide application«. A synonymous cliché is that »it will surely find practical use«.

The symmetrical stance to the forward look, also a popular favorite for conclusions, is the backward look. Take advantage of the last paragraph of your text for a recapitulation. Better yet, sum up in a couple of final sentences what you have been telling your readers. A variant on the recapitulation is the return to the **introduction**, where the concluding paragraph echoes and answers the opening **paragraph**.

Also, and this is a favorite of teachers, which they surely inherited from religious preachers, the take-home lesson. If this is your first choice, you need to coin a memorable formula for it. Or, if you find a substitute by quoting a famous person, author, text to the same effect, that will also work beautifully. Your article will close thus on a sentence by Shakespeare, Emerson or Woody Allen.

The conclusion is also the perfect location to offer a personal view, if in the preceding paragraphs you have managed to clothe yourself in a mantle of objectivity with a serene overlook.

Again, the obverse to such a stand is also worthy of note. Rather than self-centering the conclusion – if you do so, avoid the one-upmanship which goes too easily with it – instead, invite people to join you, »we encourage you to explore with us such and such«.

By offering you a taxonomy of conclusions, I may have induced you to select yet another option. In which case, not only do I applaud your imagination, I venture to predict that it will be a fitting end to your text.

REFERENCE: G. E. Kennedy (2001) *Professional and technical writing: problem-solving at work.* Prentice-Hall, Englewood Cliffs, NJ.

a hard-nosed accountant 

DEALING WITH AN ACQUISITIONS EDITOR

You are meeting this person on your campus or nearby. S/he is looking you out for having heard of you favorably by word of mouth. One of many small surprises you will experience from interaction with this editor is the awareness of strengths and weaknesses within your institution. This reflects a profound truth, great books can originate anywhere, irrespective of the worth and prestige of the institution their author is affiliated with.

The acquisitions editor is the equivalent in academia of a talent scout in professional sports. This person combines a nose for promise and knowing what sells at a given time. A nose for promise? Indeed, this person is not very much interested in what you might be expected to write about, within your specialty. Most likely, it projects into such a narrow segment of the market that it bears little commercial interest. Hence, do not expect to talk this person, who has a broad command of the whole science communication area, into buying into your field proper. His/her job, an extremely stimulating and enjoyable one, is to invent a **book**. To invent is a verb stemming from the Latin *invenire*, »to come upon, to see for the first time«. Do not rob your visitor of the professional pleasure of having elicited from you a manuscript you had no idea you had in yourself.

As I have already stressed, it may well be a book outside your field of acknowledged expertise. It may be also a book, squarely within your specialty but not addressed to your peers. The acquisitions editor believes that, from being at the leading edge of science in your particular field, you are capable of describing it to the laity. It is a top-down concept, one in which you would write for a lower readership – but of course without talking down to them.

Moreover, the acquisitions editor disguises, beneath an extremely attractive veneer of broad scientific culture, a hard-nosed accountant. This person will be negotiating with you, yes already, the number of pages of the prospective manuscript. This project has to fit within a certain budget. You will be called upon to deliver a shorter book than expected. Editors, such as copy editors in the very last stage before printing, will help you out. They will see to it that your book be better written than is your wont, or the expectation among your profession. As one such acquisitions editor is fond of declaring, »it's my job to pick up the best work I can possibly get my hands on and get it out there as well as I can.«

Think carefully before signing a contract, though. You will be committing yourself to more work, a lot more work than you expect. If you factor in the total number of hours you will be putting into this project, you will turn into a public benefactor. In terms of earning money, you might be better off hiring yourself out as a window-cleaner or as a translator. The other negative consideration is that, if indeed you accept a transdisciplinary assignment or one outside your discipline, you are going against implicit institutional rules. Only if you pull it off extremely successfully will the profession credit you with it, grudgingly. Your colleagues will act collectively in the manner of the unhappy and worried parents of a teenager who without their explicit permission goes out and does something unusual for that family.

REFERENCE: E. Goffman (1967) *Interaction ritual: essays on face-to-face behaviour.* reprint Pantheon, New York, 1982.

a hero and a villain 

DIDACTIC OR NARRATIVE?

Both, is the best answer to that question, in the relatively rare cases when a scientific writer is able to pull it off. It is a small feat, because these two modes of science communication rule out one another to a large extent. But there are exceptions. Some authors are extremely gifted, are able while **telling a story** (the entertaining part) but also to instruct (the teaching part).

Readers, for the most part, the public more generally, look forward to learning something from scientists. People often are beset by a huge inferiority complex. They remember science from school as a hard subject. They often equate it with mathematics, a subject in which they were not gifted – this is familiar ground to everyone.

Scientists are called upon to fill the void. We have to answer this intense desire for knowledge, for painless knowledge I hasten to add. Another reason with the same answer is that science can undergo swift changes. People are aware of it. They do not want to be left in the dust.

For this purpose, one has to heed some elementary rules: go slow; go step by step; use concrete examples; jettison all technical terms; recapitulate the argument at various points, and again at the very end; make sure your audience has understood the gist of your presentation. Pose some simple questions, and do not hesitate to retrace your steps.

Is this good and fine, will it suffice? Of course not. Many people do not relish returning to the classroom, if only a symbolical return. Thus telling a story, as a rule, is a better mode of science communication. Everyone loves a good story. It will enable you to inject some human interest.

Make it captivating. Many a scientist has been spellbound by the discourse of nature, as it unfolds. Thus, it might be sufficient to transcribe it, and your audience will gobble up the tales, especially if they carry the accent of natural history in particular.

Otherwise, you will have to come up with a tale. One complete with a hero and a villain, with hurdles to be surmounted by the hero or shero; not to mention a happy ending.

They say the devil is in the detail. This is the place for you to recount the actual incidents which happened during the course of the research, the personalities, the chance occurrences which either derailed the work or put it back on track.

As a storyteller, you can win the day. Easily.

REFERENCE: J. Franklin (1994) *Writing for story: craft secrets of dramatic nonfiction.* Plume, Penguin, New York NY.

embellish reality 

ILLUSTRATION

It has a threefold role: to underline the main points in the text; to go in directions the text only hints at, and last but not least, to entertain and provide visual interest and satisfaction.

Avoid redundancy. The illustration complements the text as a separate channel for information. Be imaginative. Brainstorm with whomever you choose. Come up with arresting or alluring images.

A picture is worth a thousand words according to the Chinese proverb. So true, but in different ways. If a picture is information-rich, it may save you a thousand words.

Assume that writing a thousand words takes you a couple of days: the timespan is also indicative of the care needed in selecting the right picture. Making the decision within a minute just does not wash. You need to spend much more time lining up suitable candidates, at least a dozen per space allotted. Carefully pick your favorite. Discuss your choice with other people.

Be aware that most readers will first turn to the pictures and only read the text afterwards. Accordingly, the pictures together with their captions have to tell their own story independently from the text. Put yourself in the readers' place when composing your paper. Select the pictures and devise the captions so that they present their own story effectively.

Textual illustration has a long history. It goes back to illuminated medieval manuscripts. A painter was responsible for the illuminations. Other monks copied the text. Thus, the tradition endures of separate but complementary channels.

An outstanding example of the relative independence of images and text was provided by the too-short-lived magazine *The Sciences*, published by the New York Academy of the Sciences. During Peter Brown's editorship – he now edits *Natural History* – stories in this scientific magazine were illustrated with reproductions of contemporary art making for an unexpected counterpoint which, somewhat surprisingly, worked to perfection.

Concerning magazines, the good news and the bad news is that the illustration is indeed an editorial prerogative, not yours. In the best of cases, they will run their decision by you. There are many reasons which I won't go into. Let me only say that it is best to leave the choice to the professionals. They do not dictate how you do research. Do not interfere with their job.

Which they do admirably. Pictures in magazines often embellish reality to a surprising extent. For instance, a recent issue of *The Smithsonian* carries an article on Rome. It includes a photograph of the Colosseum, which makes it look like an incredibly lyrical wonderful monument, built out of a honey-colored, even gilded stone. Whereas actually it is a rather decrepit building, grimy and grey in appearance.

You can nevertheless influence the process. Always have at the ready a portfolio of images from your research or somehow relevant to it. Present it to the editors of your piece. They are likely, if not to pick many of the illustrations from your set, to get good ideas for the iconography just from looking at the pictures you submit.

Again, write or review the captions very carefully. Give the illustration at least equal attention to the text.

REFERENCE: E.R. Tufte (1997) *Visual explanations: images and quantities.* Graphics Press, Cheshire, CT.

enjoy it 

IRONY (IN ADDRESSING THE PUBLIC)

Whether in a magazine piece, a newspaper article or a **lecture**, you the scientist, strive to be understandable, using clear and simple language. At another level you may sound lyrical or enthusiastic. Why not also sprinkle your presentation with the spice of irony?

This means distancing yourself somewhat from your message. More important, it means not taking yourself too seriously.

Do not misdirect your irony. Never address it at your readership or audience unless there is good reason for doing so. Do not talk down to people, even if what they believe in, in your opinion is misguided or just plain wrong. If you want to take issue with such beliefs, you need to do so with delicacy – which is where irony is an absolute must. Never mock your readership or audience. You may need to challenge some of their pet ideas, in which case you will counter conventional wisdom. Do it with a light touch. Leave preaching to the priesthood.

In making a presentation to your peers, you are totally committed, displaying professionalism to the utmost, with utter seriousness. When addressing the public however, some other qualities are required. You then need to be engaging, witty, funny without turning histrionic. Irony is an excellent safeguard against becoming dogmatic. What the public is yearning for is someone, a writer or speaker, who is both learned and entertaining.

If you first make fun of your own ideas, you will then be able to utter provocative and even outrageous statements. Then and only then, it is permissible to attack others.

This should come spontaneously. Laboratory life, peer review, presenting seminars tend to thicken skin. Most of us are used to having ideas and results attacked. This is how you learn to prove their worth – showing that your assumptions are sound, that the results are firm and that your logic is unimpeachable. You have also probably played devil's advocate and used a paradox if need be, in order to make a point.

In a word, irony should come easily. It is a natural product of critique, endemic within science – the attitude (which Descartes formulated in his *Discours de la méthode*) of taking nothing for granted unless it resists the strongest, the most scathing doubt.

Be aware that irony may create enemies. Not everyone is graced with a sense of humor. However, it will make you many more friends than it will antagonize people.

Polish your sense of irony. Irony should not degenerate into sarcasm, which is why it must blunt its edge. Conversely, if your irony loses its sharpness, you will be left with its little brother, **wit**. Which is good and fine – but a definite loss to everyone.

There is a way of being ironic which is truly delightful. Find it. Enjoy it as you make other people enjoy it as well.

REFERENCE: W. Zissner (1990) *On writing well: an informal guide to writing non-fiction.* 4th edn, HarperCollins, New York.

psychological need

OPENING PARAGRAPH

»One of the cherished customs of childhood is choosing up sides for a ball game. Where I grew up, we did it this way: The two chief bullies of the neighborhood would appoint themselves captains of the opposing teams, and then they would take turns picking other players. On each round, a captain would choose the most capable one (or, toward the end, the least inept) player from the pool of remaining candidates, until everyone present had been assigned to one side or the other. The aim of this ritual was to produce two evenly matched teams and, along the way to remind each of us of our precise ranking in the neighborhood pecking order. It usually worked«.

This, in its entirety, is the paragraph with which Brian Hayes starts his column on computing science in the magazine *American Scientist* of March–April 2002. It is a model, I submit, for what ought to be an opening paragraph for a magazine piece. Its foremost quality is referential. It answers right away the reluctance, to be expected from a reader, at tackling any article having to do with science – or with a specialty other than one's own. *American Scientist* with a readership chiefly of scientists, owns up to similar timidity since few scientists are generalists. We might as well directly face such reticence, which is exactly what Dr. Hayes does.

His little trick is, instead of launching directly into the subject matter of his article, »*The Easiest Hard Problem*«, i.e., the partitioning problem in number theory, starting from a real-life situation which most of us have experienced. The reader is a little bit taken aback at first, »*what on earth has this to do with science and with science communication?*« might be the slightly bemused thought, also the somewhat amused response. The reader is surprised, the very unpredictability of such an opening makes one want to proceed, to find out where it leads.

Starting with the second **sentence**, the writer goes into storytelling mode. This also is excellent. It hooks readership which, henceforth, becomes interested, truly interested. Perhaps even fascinated.

Another quality of this introduction, besides the concreteness of a familiar example, told as a simple story, is the **wit**, the **irony**. The author writes tongue-in-cheek, we can tell from cues such as »*cherished customs*« or »*ritual*«: an anthropological stance, but we know this to be folk rather than serious anthropology. We know to expect a switch to the science story immediately afterwards.

This transition into the real subject matter of the piece is embodied in the following sentence, where the author introduces his topic together with its technical name, »*the balanced number partitioning problem*«. At the end of his article, the readership has the psychological need to return to square one, Brian Hayes indeed comes back to the introductory little story, »*I take satisfaction*«, he writes »*in the thought that our ritual for choosing teams was algorithmically well-founded*«. I take satisfaction in the thought that his example may become contagious, that you too will teach yourselves to write pieces for the general public with comparable opening paragraphs.

REFERENCE: B. Hayes (2002) *The easiest Hard Problem*, Amer Sci 90(2):113–117.

the intuitive conclusion 

ORGANIZING YOUR MATERIAL

The first question to ask yourself is whether you have results which might speak to the interest of the general public. In all likelihood, the answer is negative. Your work, highly valued as it is, stands in the mainstream of a scientific subdiscipline, which does not give it wider appeal. It may well be that, as you look over a series of research articles bearing on a single problem – let's say half a dozen – none of that impressive material qualifies as fit for popularization.

Accordingly my recommendation for that initial step of identifying your material is to look over first the publications you consider marginal. Give priority to your marginal interests, i.e., topics which your expertise allows you to tackle masterfully but which, somehow, you have not pursued yet, nor plan to include amongst your professional goals.

My other recommendation is to consider seriously as potential stuff to work from anything which gave you a sleepless night. Enigmatic topics, insomnia-inducing lend themselves handily to storytelling.

Before going any further, another admonition is in order. Your mindset as a scientist is not at all keyed to the storytelling mood you have to engage in. If you are incapable of switching into this different mode, 'tis much better to drop the intent at communication.

Addressing the general public demands a lot of preparation. It is extremely time-consuming. It is hard work. It is very difficult. Perhaps, you are not the person for such an assignment. This does not reflect badly on you. You were graced with other talents than those of the storyteller. Better drop the idea than make yourself miserable and fall flat on your face – surely a most unusual occurrence for you.

You could do worse than engaging regularly in preparatory, stylistic exercises, such as taking-up daily another piece of news and writing a **paragraph** about it.

Let us assume then you are enthusiastic and well-prepared. You are about to tackle the material you have selected. There are now another three stages in the organization: (i) formatting, (ii) ordering, (iii) final selection.

The formatting part will be the key to this whole process of structuring your material, for its **necessary reconstruction** into a story. This is where you distribute your material into separate compartments, categories such as: stating the enigma; conveying the challenge presented by this puzzle; finding an analogy in ordinary language or in everyday life; whence the interest in solving this particular mystery; where the idea for a solution came from; ... At some point, you may want to relate to the standard, classical categories in storytelling. Identify the obstacles to reaching your goal, your main help in overcoming those, why the pot of gold at the end of the rainbow was desirable, the path followed, and so on.

Now you get to order this material in sequence. There will be an obvious ordering principle, logical, chronological what have you. This is, arguably, the easiest stage in this whole process.

The final selection is both crucial and excruciating. This is where you discard most of your material. The public has no taste for complexity. The scientific mind does not accept easily such rejection. Making sense of complex phenomena is, after all, what it excels at. And yet you have to find how to couch your story into a simple, easy to grasp tale. If you had a single idea to communicate, which should it be?

Let me refer briefly, in closing, to an example. *The Christian Science Monitor* published, in its December 19, 2005 issue, a piece by one of its staff journalists. Entitled »*How to explain a mini-planet's odd orbit*«, it relates the discovery of a tiny planet, in the solar system, nicknamed Buffy by the astronomers. Its orbit is peculiar for being nearly circular and for being tilted by 47 degrees off the plane of the ecliptic. The reporter focussed, mistakenly in my opinion, on the challenge to existing theories on systems of stars with orbiting planets around them.

Why is this a mistake? Because you cannot afford to go into these theories, their assumptions and their predictions. Moreover, believe me, the public has no idea of what a theory is.

It would have been better just to pursue the analogy from ordinary language: a planet out of kilter, out of whack. And/or devote more space to the working hypothesis currently entertained as an explanation of the odd findings: the Sun with its gravitational pull snatched Buffy away from another star going by. The journalist involved, Peter N. Spotts, did these three things. He might have been better off leaving alone the first which in all likelihood he was talked into by his scientific sources, all excited about the challenge to existing theories.

The take-home lesson? What ranks as number one to scientists may not even register in public perception.

REFERENCE: P. N. Spotts (2005) *Christian Science Monitor.* December 19: pp. 2 & 4.

members of the family

PERSONALIZING

In addressing the general public, whatever your format, jettison the impersonal tone you are all too familiar with from reading and writing scientific publications. Address as members of the family the people you are writing for, or talking to. Switch from the impersonal to the personal.

Why is it important, you may ask? To dispel the notion that science is foreboding and that scientists are remote and aloof. To communicate the joy of gaining new knowledge. But foremost, to remove the pulpit and make everyone comfortable. Your listeners should feel that you are not talking down to them, that you are considering them as equals.

How to do it? Since you are sharing an experience, tell your tale in the first person. Remain modest. You may do worse than recounting, as a part of your narrative, the incidents: how you were sidetracked, how you made a mistake or a blunder, how you learned from it and nevertheless went forward.

In other **words**, use as a model an explorer's description of difficult travel in an exotic land. Help your audience identify with you through the troubles you went through. They will be all the more ready to admire how tenacious and smart you were in devising a solution to reach your goal.

Convey to people what life in the laboratory is like. In particular, show them how discoveries can and do arise from small observations, of the »God is in the details« type.

Put questions to them, to draw them inside your story. At this stage, what would you have done? What is your guess as to the outcome of this particular experiment? What is the intuitive **conclusion** one would have drawn from these results? What did I forget, in your opinion? Can you remind me of what my starting assumptions were? What would have been your conjecture, at that stage?

Tell anecdotes. People love anecdotes – just as they relish being told a good joke. As a rule, the anecdote is the very last thing people will forget.

As an example, take Newton's apple story. It has no basis whatsoever in history. It entirely lacks documentary evidence. In all likelihood it was a late invention, much posterior to Newton's lifetime.

Is it then entirely devoid of merit? Yes, to the historian of science. But, to the general public, the answer has to be a qualified »no«. The anecdote has merit on three counts. It reminds people that a chief contribution of

Newton was to formulate the law of general gravitation. It tells them that as a child he became interested in physical phenomena and their explanations. And it brings this all-too-serious and even solemn British gentleman down to earth.

In like manner, your readership or audience will relish your narrative if peppered with a few anecdotes regarding events which occurred in the course of an investigation, or that you witnessed.

REFERENCE: M. Turner (1998) *The literary mind.* Oxford University Press, New York.

willingness to learn 

RESPECT

This is an essential element of science communication. Whatever message we prepare for the public as scientists, we should be careful not to talk down to our audience. We have to refrain from the paternalistic, i.e., from an attitude of »we know better, let me tell you how things are«. The reason for avoiding such arrogance is not so much that it is all too readily perceived and hence resented by the recipients. It is, first and foremost, that we operate in a democracy. Our fellow-citizens are intellectually our equals. Such an axiom – whether it is an aporia is morally irrelevant – is basic to any attempt at science communication. The brotherhood of mankind is our governing principle as science communicators.

In other words, the scientist addressing the public, whatever the channel, must do so with respect. This is not an artificial attitude, one required by social conventions or for selfish reasons of self-protection. There are sound reasons making the public, this vague social entity, inherently respectable.

Its vivid interest in science is one. It is most endearing on the part of people to go beyond their immediate needs, their material well-being, and to show curiosity for what we discover about the world, about the behavior of things. This is a call we ought to answer with utter seriousness, sincerity and without ado.

The willingness to learn is another admirable feature in the attitude of the public towards science. We should be all the more appreciative that it coincides with our own. We too are students for a lifetime. That our fellow-citizenry, obviously not continually but epidodically for individuals, is willing to keep learning, is deeply moving. It ought to make us into more than willing teachers. Respect your virtual audience for the strength of its desire.

Such desire goes with suspension of disbelief. To a large extent, science has already replaced religion in the public mind, as an ideology, a set of ideas to be looked after and to be revered. Can we not show respect for the people who thus put us on a pedestal?

The main argument, though, to induce such respect on our part is the aggregated experience of our public. The sum total of the knowledge, which any of those individuals possesses, dwarfs ours and it becomes akin to a wisdom. In front of it, we cannot but find ourselves awed and humbled.

REFERENCE: J. Flaherty (2005) *Coaching, evoking excellence in others.* Elsevier, Amsterdam.

guilt of obese people 

RHETORIC

In Antiquity, rhetoric was a major subject of study. Any public figure, the politician, the philosopher, the attorney was steeped in it. Rhetoric was the art of convincing listeners to a speech, independent of the subject matter.

At this point, you may well ask, what has this to do with me? It has to do with the current unease with science, to put it mildly. A hook is thus needed to draw the public to any science topic. Rhetoric supplies this piece of gadgetry. I cover here some standard devices. You can pick whatever format works best for you – or come up with your own, entirely different. These are devices meant for stories, for pieces in the media such as a **magazine article**, a radio presentation, a continuity for a **television show**. Their joint feature is simple-mindedness. These formats are so crude as to be embarrassing – yet they work.

Answering an Apparently Naïve Question comes first. A highly popular French radio show, »Les P'tits Bateaux«, follows it. Children, ranging in age from about six to 10, phone-in questions on extremely diverse science topics. They are submitted to a panel of scientists. One of us chooses to answer a particular question. Two minutes is the air time allotted, as an interview with the show hostess, Ms. Noëlle Bréham. Even though the questions originate with young children, the listeners are predominantly adults. This particular rhetorical ploy, the Naïve Question, raises difficult scientific problems, such as why the sky is blue (John Tyndall and Lord Rayleigh), or why is there night and day (Olbers's Paradox). I used it myself, not long ago, in writing a little book entitled *Why Is the Sea Blue?*

Another popular rhetorical format is the Debunking of a Commonly-Held View. As an example of use by the media is an American cable TV channel broadcast (autumn 2004) entitled »Mythbusters«. The game is to demolish a commonly-held opinion, on the strength of scientific facts.

It might be, for instance, as in a magazine article published in 2004, the notion of the guilt of obese people. Are they not overeating? Can they not stick to a healthy diet, lose weight and stay slim? The debunking of such commonly-held beliefs argues from statistics that obesity has a strong genetic component. It conjectures that this genetic trait, affecting a segment of the population, to have had survival value in the distant past for hunters-gatherers, in contrast to agriculturalists.

I shall mention now the Vindicated Thinker, the hero who surmounts various hurdles to reach a goal. People like to identify with a person persistent in the face of adversity. It gives scientists a more human face. There is nothing like a tale of woe to excite sympathy, hence interest. Hindrance to the hero's progress can be self-inflicted, as in *A Beautiful Mind*, the book on John Nash and the movie starring Russell Crowe.

Related formats are the Daring Pioneer and the Visionary, also in the form of biographical portraiture. Stories about Alfred Wegener, of continental drift fame, or about Alan Turing, who theorized computation at an early stage, are usually couched in the former; whereas stories about Albert Einstein or Stephen Hawking often use the latter kind of rhetoric.

Hero-worship takes many forms. Within the oral tradition in our profession, it is often the Bigger-Than-Life Person. This can also be turned into a good stereotypic story format. Two related clichéd narrative modes, at opposite poles in readers' sympathy, are the Scientific Gentleman/Lady and the Mean Curmudgeon.

The last of such hackneyed, but efficient formats I'll mention here is The Controversy, which seldom fails to make the public attentive. Of course, it can be hybridized with any of the above. For instance, you may combine The Controversy, the Mean Curmudgeon and the Visionary.

An engaging paradox pits the sophistication in any rhetoric against the above themes being simplistic, to say the least. The answer, which it is not the place to elaborate upon, is the force of popular culture (Hollywood movies, best-selling books, …). Our aim is to make scientific culture an integral part of culture. That it should infuse pop culture too is self-evident. It is a moral obligation of the times that scientific highbrow be turned into lowbrow entertainment. We hewe to that line.

REFERENCE: E. P. J. Corbett and R. J. Connors (1999) *Classical rhetoric for the modern student.* 4th edn, Oxford University Press, Oxford.

a chatty, earthy manner 

SEDUCTION

Faced with a public whose attitude to science is at best a lukewarm mix, made of extreme curiosity together with some fear and an element of rejection, you want to show it with a smiling face. You need to first seduce your readers or your listeners so that your message can get through. It would be a pity if it met with such resistance that people would not even consider it. Hence the need for seduction.

How can you do it? First and foremost, by your storytelling. Yes, it is a skill. Yes, it can be improved. How? By practice. Write one-paragraph-long short stories. Choose bizarre incidents which strike your fancy, tell them in a pithy way and read them aloud. This will help you hone this particular skill, essential to any kind of science communication.

One way in which you can seduce an audience or a readership is by starting your presentation with some **anecdotal** material. An anecdote or two, told in a chatty, earthy manner will convince your public that you are a good communicator, that you won't go over their heads with material that is dry, remote and foreboding.

With such a start you will convince people that your language is simple, that you are capable of a direct and modest tone, that your purpose, far from one-upmanship is communication, that you are intent in sharing your knowledge with your fellow-citizens.

»To seduce« stems from a verb in the Latin language, *seducere*. It means originally »to lead astray, to mislead«, which is totally at cross-purpose with what you are aiming at. Be aware that the down side of a seduction is that people are wary of the seducer. You do not want your message which is very much genuine, to sound like a con job. You do not want it to be construed as an attempt to tell a bunch of lies. Accordingly, you will have to stay in the shadow of your presentation, not make any attempt to step in front of it and take first place.

Which does not mean that you need to make yourself into a bland personality. Not at all. People, even though suspicious of an attempt at seduction, love being charmed. And this is where your personality comes in.

Express it, find your authentic voice and tone. Be original. If you have the ability, make transdisciplinary connections. Surprise people with your intellectual agility. Show them how much a scientist can differ from the stereotype. They half-expect a Dr. Frankenstein or a Dr. Strangelove. Show them instead a nice and warm fellow who is humorous, who can hold them spellbound with fascinating stories, and who has worked hard at being an effective communicator.

REFERENCE: E. Goffman (1980) *Relations in public: microstudies of the public order.* HarperCollins, New York.

watch performing artists 

SPEECH DELIVERY

Practice makes perfect. Nowhere is this more true than in the intricate art of public speech. Experience makes people good at it. To mention professions for which the vocal dimension is foremost, actors and teachers usually excel at it.

Rehearse often your presentation, not only isolated segments but also at least one nonstop run-through.

There is a great deal you can do to improve your own performance. First, listen to yourself. Watch a video of your speaking or even reading aloud. You will be your own best critic. Note the defects and the annoying mannerisms. Return in front of the camera for an improved delivery.

There are pre-requisites to good speech. Set your body properly. Be dressed comfortably. Leave your neck free. Your chest and your feet should likewise feel unconstrained and of course painless. Discover which resting positions work best for you. Arms closed across your chest? One hand in a pocket, or posed on the lectern? Make a repertory of at least three such stands between which to alternate. Crucial is the anchoring of your whole body onto a »support«: for instance, a pencil held in your hand or between your hands can serve as such.

You do not have to remain static, to the contrary. Watch performing artists on a stage. They keep moving, don't they? Follow suit. Find which motions come to you naturally. Obey the urge and espouse their rhythm.

You are almost ready to take the floor. Breathe deeply, make several conscious inspirations and expirations. They will help oxygenate your brain adequately, and keep you alert. Start speaking. Memorize and perhaps even rehearse, to someone, or to your mirror, your first lines. They will thus come easily to you.

Try not to read your talk. It will sound much better if seemingly spontaneous. You can help with cue cards to look at from time to time. Your audience will respect your need for an occasional pause, and for taking a breath – literally. They need it too.

While speaking, do not forget to breathe. Make your **sentences** short enough that you can utter each in a breath. Pace yourself. A public speaker is like a horse running on a track: once a pace has been chosen, to change it is tricky. Make sure to talk conversationally, neither rushing through the words nor being agonizingly slow. Study your written text sentence by sentence and mark it. Your breathing needs being brought in harmony with the punctuation. Your delivery will have to convey it by itself.

Volume? You want to be heard throughout the room. Avoid being too loud. Exercises will teach you to project your voice. For instance, say in succession the same word or a sentence, such as »can you hear me?« in order to be heard at distances of one, five and twenty meters. During your talk, make sure to vary occasionally the volume of your delivery, to avoid monotony. For emphasis, you may want to say some of the things in a tone of confidence, which goes with a softer voice – a trick which every teacher uses on a rowdy room needing to be silenced and to become attentive.

Likewise, give different colors to your voice. An extremely useful exercise is to repeat a single word or sentence, such as »condensation of droplets«, in various tones, those of assertion, logic, anger, sharing a secret, being lewd, etc. Such exercises are best performed with a tutor. At times, you may want to emphasize a single word: pronounce it slowly, over-articulate and detach syllables from one another: con-den-sa-tion.

Pitch? It is important that your voice be pitched at its natural, strain-free level. Learn how by reading a text aloud. If you are a beginner, do such an exercise five minutes twice daily. Singing, if you are so gifted, will help you greatly. Being tutored by a professional works wonders. Watching your audience – your audience, not the projection screen – scanning their faces and making eye contact, these will help you to find and maintain your pitch.

In so doing, find your own voice. This is the basic secret for seducing your audience. If you make people feel there is a genuine person talking to them and sharing knowledge with them, they will bear with you to an amazing extent.

Some people are incapable of it. They speak in an affected, unnatural voice. This unfortunate personality trait probably dates back to teenage years, or even earlier. Their assumed voice is part of their mask, of their public persona. Any audience detests such phonies. The very word »phony« says it all. If you are unfortunate enough to be such a person, you will have to undergo a major change and come out of your hiding place. It may well be that only a psychotherapy will do.

Such a pathology serves to show how fundamental a voice is, to interpersonal relationships and thereby to being an efficient communicator.

REFERENCE: M. Platow (2002) *Giving presentations in the behavioral sciences and related fields. A practical guide for the novice, the nervous, and the nonchalant.* Psychology Press, Taylor and Francis, Hove UK.

Having an attentive audience 

TAKING THE FLOOR

Addressing a lay audience is a challenge. It is an honor too. Quite a few people in the public see a scientist for the first time.

Remain silent awhile after you have been introduced. Take a good look at the audience prior to opening your mouth. The anticipation makes your words more desirable. Be aware of your **body language**. Make sure to project it at your friendliest. In taking the floor, you are reaching out to strangers. Your whole demeanor and your first words have to acknowledge this situation. You have been picked as an ambassador for the Republic of science.

Make sure you are being heard throughout the room. If need be, lower your voice to induce listeners to become attentive.

Dispense with notes. Do not read the text of your lecture. Instead, talk in a natural way. Have recourse to a few cue cards if need be. Other pitfalls include talking down to the people and using strange words. Choose your words with care and deliberation.

Admiration from your listeners is almost a given. Don't destroy it. Avoid coming across as a self-important person. Everyone is allergic to pomposity and bombast. But your worst possible mistake is to be boring.

As you take the floor, as you are about to start talking, it will be very helpful to focus on a mental image. Mull on it in advance. You need to picture to yourself a symbolic equivalent to what you are about to do. The image I favor, personally, has me walking across a river on a bridge. It connects the scientific community, on one side of the river, and the polity on the other side. I am in the process of crossing that bridge. I have nearly reached the city on the other side.

(As for the technical aspects of acting as a speaker, all the recommendations in the other entry of the same title are applicable).

REFERENCE: S. E. Lucas (1983) *The art of public speaking.* Random House, New York.

spinning your yarn 

TELLING A STORY

It is a joy to tell people what we do. There is no better way in so doing than storytelling. People love to be told stories. Having an attentive audience will please you as well. Right?

The best way to do it is to derive the story from your own scientific results. It is best to pick an episode in which you conducted science in the problem-solving mode. This will be easy, since much of science conforms to such a description.

Proceed first with the problem. How did you become aware of it? What or whom alerted you to it? What led you to address it? Did you have an intuition as to what the solution might be? What made it an important problem to solve? What was the nagging force that kept you awake at night? How did you obtain funding to carry out the investigation?

Now tell the tale of how you gave the problem a solution. In spinning your yarn, an element of suspense is needed. It can be a reminder of the puzzle you are trying to solve. It can be an unexpected finding, an incident in the laboratory, a temporary derailing of the investigation. Keep the public a little breathless by stressing repeatedly the importance of the puzzle you have addressed and by using the trick of the thriller-writer, that is, putting the hero of the story in jeopardy at the end of each episode. If in a quandary about the sequence of the episodes, just use the chronological order.

To start the story, after you have set the problem to be solved, ask yourself the question of the story teller: who are the characters in the cast? Do not feel constrained to just human characters. Some of the protagonists may be components in the physical system under study.

What were the obstacles you had to vanquish in order to solve your problem? Recount the disaster-prone part of the story. Devote one episode to each hurdle. Describe who or what gave you help, and what the key was to successfully overcoming this particular challenge.

What was the main obstacle? How did the right idea come to you? Were you able to scale that peak or bypass it?

Try to introduce a human element throughout your narrative. It will grab the interest of your readership, or of the people listening to your tale.

Your story cannot end just with the successful solving of the puzzle. It needs a take-home lesson, served in as pithy a manner as possible.

Do not feel intimidated by the task, it is straightforward. Tell what happened, as you might to a child. You may wish to use a model to guide you along. The style of any whodunit will do fine. Or, before putting pen to paper, read a short story by one of the masters of the genre such as: John O'Hara – Joyce Carol Oates – Jorge Luis Borges – Alice Munro.

You can be assured in advance of a receptive and extremely grateful audience.

REFERENCE: P. Rubie (1996) *The elements of storytelling.* Wiley, Hoboken, NJ.

equal to the task 

THEMATIC VARIATIONS

Hone your writing skills by composing variations on a theme. Nothing could be simpler. Take a few **sentences** out of one of your professional journals, and transpose them into a **paragraph** for a general readership.

Here is an example. First, the seed, from a genuine scientific paper:

»Ubiquitination is a rapid and reversible post-translational modification that is involved in numerous aspects of biology. The 76-residue polypeptide ubiquitin fulfills essential functions in eukaryotes through its covalent attachment to other intracellular proteins. Historically, the best characterized role for this modification is the targeting of proteins for degradation by the 26S proteasome after the transfer of an ubiquitin chain of at least four units, referred to as polyubiquitnation.«

Focus on just a few features, maybe a single one. Now write about it:

»Ubiquitin, thus named for its ubiquitous presence in higher organisms, is a biopolymer stringing linearly 76 aminoacids. Ubiquitin attaches itself to proteins, also formed of polypeptide chains of amino acids. Once tagged by ubiquitin, a protein is marked for the slaughterhouse. It will undergo destruction at the hands of a complex piece of cellular machinery, the proteasome.«

And accumulate a few such variations, in the manner of Raymond Queneau with his *Exercices de style*:

»Once parasitized by the small protein, ubiquitin, a protein in the body bears a marker. This induces a number of cellular processes, such as degradation and removal of the targeted protein«.

Here is another:

»To any cellular protein, embrace by its smaller congener, ubiquitin, may be the kiss of death. It is the signal for its destruction within an entity known as the proteasome«.

And yet another:

> *»Does a cell look askance at a protein labeled from attachment of the small ubiquitin molecule? It would seem so, since the protein may now be marked for the death row – filing for disposal in the death chamber known as the proteasome«.*

The point of the exercise is to demonstrate to yourself that your imagination is more than equal to the task. There is feedback too. By addressing a general readership, if only fictively, you improve the quality of the writing for your peers. Some of the above variations, not only might come as handy substitutes for the original text (the »theme«), they rephrase it in focussed mode, one idea per sentence.

REFERENCE: J. Perroy, S. Pontier, P.G. Charest, M. Aubry, and M. Bouvier (2004) *Nature Methods* 1(3):203–208.

a naive misconception 

THE NECESSARY RECONSTRUCTION

Because it is so widespread, one has to dispel a naive misconception. According to that view, the only talent needed for communicating science to the public is the ability to translate technical terms into words of everyday language. Indeed, simple and clear expression is required. If a necessity, it is nevertheless not sufficient.

To remove the technical wrappings surrounding knowledge does not magically turn it into an item for communication. There are other major obstacles to diffusion of knowledge outside the sphere of its origins. They are not removed by recourse to an interpreter, who will translate from scientific discourse into common language. The notion of a Third Man, of the popularizer as translator, is a myth. Too often it is a mere screen for intellectual laziness, even more on the part of the emitter than on the part of the receiver of the message.

For a message to get through, prior reconstruction is demanded. In order to induce and satisfy the interest of your public, you need to steer clear of the style of the initial message, that to fellow-scientists. If it was couched in legalistic, argumentative, polemical, assertive or simply didactic terms, you have to renounce it in favor of a simple narrative. Put another way, you should move from the objective to the subjective.

This is not the place to elaborate, but didacticism is the major culprit in the production of popularization which is mediocre and which fails to hit the target. Narration is the most-traveled avenue for effective knowledge-sharing. »Imagine yourself explaining your work to your grandmother« is an excellent rule of thumb.

Indeed, it is more important to transmit the thrust of living science, its systematic self-doubt, the wonders of both nature and of making sense of it, than bits of knowledge, ephemereal by their very nature. Moreover, they lose most of their meaning when removed from their context.

How then should you perform this indispensable reconstruction? To present your results to your peers in a **seminar** or in a journal, one can make use of logical argument. This won't wash for communicating with the public. Drama is your novel mode of expression.

Scientific publication is couched in the manner of a legal brief: numerical tables and graphs are the equivalent of the exhibits which attorneys display in court. Turn now to a totally different presentation. Tell a story.

How should you do that? There are various possibilities. You can choose a »show and tell« procedure. Use for instance illustrations to present the phenomenon you have studied, the apparatus you have used and your conclusions.

In any case, you will be telling a story. Structure it as a sequence of episodes. Devise a cast of characters. Turn the true history of the investigation into a plot, with twists and turns prior to a satisfactory resolution. Inject a forward impulse, making your readership and audience avid for more.

Again, switch from the objective to the subjective. Jettison the didactic in favor of a narrative.

REFERENCE: J. Franklin (1994) *Writing for story: craft secrets of dramatic nonfiction.* Plume, New York.

a friendly voice 

TITLE OF A BOOK

Any book of science popularization hooks its readers with the title. This may seem self-evident, but it will justify your effort in coming-up with a good title. And yes, the title comes first. A book cannot be written unless you know what to call it, at least tentatively.

Best-selling books almost always have outstanding titles. Some classics of science popularization are James D. Watson's *The Double Helix*, George Gamow's *One, Two, Three ... Infinity*, Stephen Jay Gould's *The Panda's Thumb*, Richard Dawkins's *The Selfish Gene*, Steven Weinberg's *The First Three Minutes*. The lesson here is for the title to be the pithiest summary, the most economical condensate one can make of the contents of a book. The title is the message, it is as simple as that. Sometimes, a book will sell on the strength of its title exclusively, whether people actually read it or not. Stephen Hawking's *a Brief History of Time* is an example.

That a title is a message can be proved. Consider only the progeny of Weinberg's classic title, such as *The Last Three Minutes* by P. C. W. Davies and Paul Davies, or *Just Six Numbers. The Deep Forces That Shape the Universe* by Martin Rees.

What are the ingredients of a good title? Brevity, originality, euphony and the typography. Let us consider each of these elements – none of which a total obligation – in turn.

Brevity is required by booksellers. Assuming that you achieve a paperback edition, the title is the hook for the attention of a prospective buyer glancing at the cover. It is a sales pitch, about the shortest one can think of, a kind of a primal scream: »buy me, I'm fascinating.«

Originality? Any title connects the contents of the book with the mental world of the reader. The book deals with a foreboding subject matter, science. Its title emphasizes a friendly voice, an easy read, a writer capable of being humorous, someone who will address the common concern and wax metaphysical, but in the simplest of terms, *The First Three Minutes* or *A Brief History of Time*. A successful title is such a tongue-in-cheek expression, it conveys wit on the part of the writer.

»Science of Mind« is not a good title for a book about cognitive science, it is too ordinary. Neither is »Cognitive Science«. Titles of scientific disciplines, as a rule, make boring titles: »Oceanography« or »Nanoscience«. *Nanorobots* would be a far better title for the latter.

The word »euphony« means »to sound right«. A good title should be said aloud easily. Tongue twisters won't work. Avoid disastrous near-homophones and puns.

The typography is important too. Coming back to my hypothetical example, it would be better to spell it *Nano-Robots,* for maximum impact.

What about a subtitle? This depends on the fashion at the time of publication. Currently, subtitles are in vogue. They are descriptive labels, to some extent first sentences in the book drawing in the prospective reader and buyer.

The procedure to follow is what parents do when choosing a name for a forthcoming baby. They make lists and they rank the items. They negotiate.

Put together a list of a dozen titles. Put it in a drawer. Come back to it after a day or two, cross out all the nonfunctional entries, add a few others which have come to your mind. When you are down to two or three strong proposals, discuss them with your editor and with the sales people.

REFERENCE: L. S. King (1991) *Why not say it clearly: a guide to expositiory writing.* 2nd edn, Little, Brown & Co, Boston.

transparently honest

TITLE OF A STORY

A good title is appetizing, it creates interest on the part of a prospective reader for an initial attempt to enter your story. It will be up to you then to follow-up on the promise of the title.

There are various ways in which to play this particular seduction scene. You can show to the reader, with the very title, that you speak the same language, he need not worry about the difficulty of the piece. *The Stuff of Genes* enacts such a promise. Likewise, *Bugs in the Brain*, with the additional merit of the B-B alliteration.

By being transparently honest, announcing the crux of the argument and conclusions in the title, you will attract quite a few readers too. Titles of **magazine articles** published in 2004 which thus served as a preview of the conclusion of the article, included: *A War on Obesity, Not the Obese;* also, *In Click Languages, an Echo of the Tongues of the Ancients*; and *A Prolific Genghis Khan, It Seems, Helped People the World*. Obviously, such conclusion-preview titles run long, they amount basically to a whole **sen-**

tence, which can work. Notice though that they carefully abstain from use of complicated words, Genghis Khan helped to people, not to populate, the world.

A whodunit title is great, such as *Crows: A Murder Mystery*. Surrealistic titles are even better: *The Fish That Swim in my Head*, also from a 2004 magazine article. Funny and thereby enticing titles pun on recent events in popular culture, such as the titles of successful movies. Two such titles in 2004 were *The Silence of the Lambswool Cardigan*, a take on the *Silence of the Lambs*, and *Good Whale Hunting*, punning on the movie *Good Will Hunting*. Surely, similar expressions are current in your own lab talk, referring to various features in your research, tongue-in-cheek and self-deprecatingly. I bet you they would make superb titles for magazine articles.

Most titles, about a third of those for science pieces in magazines I'd guesstimate, consist of questions, rhetorical questions to be more precise. Two magazine science pieces in 2004, by different authors I hasten to say (meeting of minds), were entitled *Are Women Necessary?* and *Are Men Necessary?* An article analysing statistics on first names and the fashions influencing them, from data posted on a Web site, was entitled *Where Have All the Lisas Gone?*

The rhetorical question is all the better if it offends common sense or conventional wisdom. Such titles include *Can We Trust Research Done with Lab Mice?*, *Is War Our Biological Destiny?*, and *What Happened Before the Big Bang?* This last title challenges the by now well-accepted cosmology, by which the universe originated with a Big Bang. Hence, to ask what happened before the origin of the universe, an apparently meaningless question, ought to stimulate curiosity. Likewise, an article on genetically-modified organisms was entitled *Will Frankenfood Save the Planet?* Here again, the apparent contradiction between Frankenfood, a neologism and a portmanteau word (two meanings packed into the same word) coined from »Frankenstein« and »food«, and the notion of saving humanity from starvation and extinction, is enticing.

This brief typology surely will have given you ideas. Now you know, even if you do not have definite plans to write a magazine article soon, how to entitle it. Your imagination, trust me, will work wonders.

REFERENCE: E. Hancock (2005) *Ideas into words: mastering the craft of science writing.* Johns Hopkins University Press, Baltimore MD.

pithy prose 

VOCABULARY

(IN PIECES WRITTEN FOR THE GENERAL PUBLIC)

Use of the accurate and terminologically-correct word makes the vocabulary of science stereotyped and repetitive. While true of our professional writings, this limitation needs not extend to science popularization. As soon as one jettisons the translation fallacy, i.e., the mistaken notion that all it takes is to translate from the formal idiom of science into everyday language, the available vocabulary expands immeasurably.

The vocabulary of narrative has considerable leeway and range. There are many ways in which to convince yourself of the truth of this statement. Read for instance a historical work.

What works best, I will argue, is pithy prose of simple **words** and short declarative **sentences**, occasionally enlivened by the apt recondite word. Provide gratification to your reader from infrequent use of a sophisticated word. Hemingway-like prose works beautifully for a series of actions. However, science is more reflexive than the everyday language we use. Here and there, you need to put in a rich word, so that the reader does not doze off and so to offer the satisfactory feelings of not being talked down to and of learning a few things besides.

Peruse the dictionary. You may want to start compiling a thesaurus of words you like. They will come in handy, sooner rather than later. Open a dictionary at any page. Its riches are there for the taking.

I have just done so, and looked up derivatives of the simple word »fish«. Here are just four, culled from about five times as many. I am convinced I will soon find ways of using them in my writing. The *fish story*, this is too obvious to elaborate on, is a tall tale. *Fishybacking* is the way in which loaded truck trailers and containers are transported, on a barge for instance. A *fisheye* is a surface defect in the shape of a spot. And a *fish-bellied* object sports a convex underside. Those four words are self-sufficient. You do not need to explain what they mean, the context will make it clear. Moreover, and I beg to be forgiven for repeating it, you will gratify your readers by treating them to words which they did not know. Moreover, they are not tongue-twisters nor are they at all opaque. They may be unfamiliar, but they are uncomplicated and easy to remember.

Do not shy from the archaic. No one uses but everyone understands the verb »to eschew«. It means »to avoid, to abstain from«. When would you use it? While it has no place in a straightforward story (do not write »Susan eschewed ordering dessert«, which would be ludicrous, write instead »Susan avoided ordering dessert«), this fully-blooded word may take a choice place, for instance, in a discussion of an epistemological issue (»Dr. Roberts eschewed the deductive in favor of the inductive inference«, where it will sound natural, just like a William Morris fabric in Victorian era furnishings).

Likewise, the lexicon from the professions and crafts will provide you with a treasure trove of words with which to enrich your vocabulary, and thus enchant your readers. One of the main attractions of whodunit novels, with the pathologist or the archeologist as combination hero and detective, is being invited to sample, in context, technical names of bones, such as the humerus, the femur, the scapulum, the nasal sill, the hyoid or the malars.

You may differ: to me, a good meal is incomplete without cheese and wine, not to mention dessert. A good read is likewise incomplete without a few rich words.

We should note, however, in closing this entry that the scientific community now consists of a majority of non-native speakers of English. Those colleagues, whose work is also high-quality and crucial to the advancement of knowledge, may be unable to enjoy reading a scientific paper with a few rich, i.e., unknown words. Be kind to them: if indeed you sprinkle your prose and spice it with the occasional rare word, make sure that its meaning be obvious from the context.

REFERENCE: L. Dupré (1998) *Bugs in writing: a guide to debugging your prose.* revised edn, Addison-Wesley, Boston MA.

PART II

THE GENERAL PUBLIC

GENRES

AUTOBIOGRAPHY

This is a difficult and treacherous genre. Writing about oneself runs the risk of narcissism and of inflated self-importance. Inaccuracy, as in scientific work, is a cardinal sin. These are the negatives.

But there are so many aspects in an autobiography, most attractive to readers. Young people may find a model figure from it; many a calling to science has thus been seeded. This motivational appeal is crucial. Your **book**, furthermore, will offer insights nowhere else to be found. It will put results in context. It will give science a human face. It will sweep away the prejudice painting it as remote and haughty. Anecdotes are not to be shied from, for they convey life in the laboratory, and the human interactions behind any research result.

Accordingly, in line with public demand for biographies and autobiographies, which as a rule sell very well, should you start writing yours? You ought to be aware though of a shadow over the genre as a whole. It is the reticence of historians. As a class, they are highly suspicious of such documents. Why? Because they are trained to hunt for facts from multiple, cross-checked sources. Because they prefer indirect evidence, where there is less likelihood of a self-serving account. Basically, anything coming from the horse's mouth, they are wary of.

Prove them wrong. Make sure that you have outside corroboration for any event or episode you write down from memory. Another reason for the jaundiced judgment by historians is that they do not trust the participants with the objective, bird's eye view, which in their opinion should prevail.

Moreover, autobiographies, they say, tend to not only exaggerate the importance of the writer but, more generally, that of the human actors. They tend to stress the role of individuals and neglect or soft-pedal that of social forces, collective actors, or mere objects and instruments. The current buzzword among both historians and sociologists of science is context. An autobiographical text has value for them only if it presents the context (material – social – historical – etc.) for a life, a piece of work, or any achievement.

Pretty scary, isn't it? Some colleagues, few fortunately, are thus brainwashed into inaction. Others, while not necessarily intimidated, for various other reasons, the amount of work involved, unease at writing and at writing about themselves, also demure. Quite a few people prefer to keep this kind of writing private, at least within the confines of their family.

But if you feel the urge to write your recollections, just DO IT. My recommendation is that you seek a publisher. Collaborative work with an editor can be most gratifying. Likewise, you will be most capably assisted by the detailed response from a number of independent and capable reviewers.

Besides the likely hit with the general public, the worth of a text of yours to historians is the representative character of your own trajectory. It will have value if it supplements textual evidence, if it provides information otherwise absent from written documents, whether publications or archived correspondences. However, refrain from hearsay. Report exclusively what you have witnessed. Give the names of people who were also there and may have also recorded the event and its circumstances.

Concentrate on the very few episodes, encounters, chance happenings, … which turned out, from hindsight of course, to have been determining, in one way or another. Try to capture what is most elusive to historians and therefore likely to be very useful to them: a *mentalité*, i.e., a collective state of mind at a given time within the profession, or within a restricted group of peers; what people were talking about in the labs – not gossip, but the small talk that at any given time reflects it; the life cycle of an idea, good or bad; the small auxiliary tools in the daily life, such as the Xerox machine, the Polaroid camera, the tape recorder, …; portraits of great scientists who, for one reason or another, have remained unknown or unrecognized.

However, an autobiography is not the place to settle scores. You may be eager to denounce operators and rogues, people having stolen ideas from others or having otherwise behaved badly. Don't do it. You will appear as a mean-spirited old curmudgeon, which is probably not how you wish to be remembered.

In short, you as an individual are truly unique. You are most probably capable of the autobiographical effort. What will make it original and worthwhile is the combination of first-hand experience and of your personal take on people and events.

REFERENCES: F. A. Carroll and J. I. Seeman (2001) *Placing science into its human context: using scientific autobiography to teach chemistry.* J Chem Ed 78(12):1618–1622; Bernard Selling (1990) *Writing from within: a guide to creativity and life story writing.* Hunter House, Alameda CA.

what made a person unique

BIOGRAPHY

With this genre, one narrates the life and times of an interesting scientist. The cottage industry of biographies of Galileo, Newton, Darwin or Einstein has perhaps little need of additional entries. Conversely, there are quite a few scientists of the first rank such as Johann-Heinrich Lambert, the Franco-German philosopher-cum-scientist, for whom a good biography is highly desirable.

With the above working definition of a biography, it helps to spell out the differences between a biography and an **autobiography**. Each genre has its merits. And each genre has its blind spots. An autobiography has an element of the confessional. It mixes reticence and divulgation of one's true self. A good autobiography conveys the feelings of the author.

Conversely, a biography of Dr. X tells us what kind of a person he or she was, which an autobiography can fail to do. Normally, people do not have the necessary remoteness to place themselves at all accurately.

A biography is that part of science history which focusses near-exclusively – as a rule, too exclusively – on the human actors, in contrast to forces, such as material factors, technological advances, social pressures and the like. An autobiography relishes funny stories, folk tales which scientists exchange when they meet, and which they build into a mythology. Its downside is the risk of becoming exclusively **anecdotal**. It is important for scientists, who relish telling one another such stories, that while their presence may strengthen a biography, they cannot serve as its mainframe.

To write a biography is an attempt at describing what made a person unique. You cannot write a good biography if you cannot stand the individual portrayed. You cannot write a good biography if your admiration gets the better of you and smothers your critical sense. This is important. You need first to espouse a stance of identification with your subject, followed by the opposite, an aloofness, when you seemingly distance yourself from Dr. X. Not that you will manage an impossible objectivity. Do not become so much infused with the personality that your biography only parrots Dr. X's views.

Your biography also needs to be critical. In passing judgment, however, you will have to refrain from the Seven Sins of the biographer, viz. 1. pettiness; 2. hagiography; 3. a Whig perspective; 4. superficiality; 5. the twin synecdoches of mistaking the person for the field or vice-versa; 6. becoming lost in the details; 7. mistaken interpretations.

Pettiness: always resist the temptation to shine at your subject's expense. Hagiography: the obverse temptation is to set the subject on a pedestal. It has to be resisted, too. Whig perspective: this defect, general to scientists interested in the history of their field, is to interpret the past in the light of the present. Superficiality: in performing a public service, your study must dig deep, or become useless. A thorough study is a better read, too. Be wary of hearsay. Try to get the facts preferably through a written document. Better yet, several such texts will enable you to cross-check sources. Synedoches: you need to present a fellow-scientist in context, neither the foreground nor the background should be allowed to steal the show. This is a common pitfall. It is always hard to balance between including enough contextual material and overemphasizing the background. Details as a maze: self-explanatory. Mistaken interpretations: use the same standards as in your science, any conjecture has to be tested against the available evidence.

Having an urge to write the biography of Dr. X means that, besides your fascination with this person and his or her accomplishments, you bring some special insight to the work. Spend at least two or three years just gathering the necessary documentation. This is not just collecting material. In this initial phase talk to people, start writing and narrowing your project, get to know similarly engaged people, start networking. Learn for yourself the field in which your subject was active. Become totally familiar with the papers published, the controversies, the correspondence with contemporary fellow-scientists, the laboratory notebooks. Acquire a working knowledge of the various contexts Dr. X was operating within, historical, social, cultural, academic or industrial, … and situate Dr. X's contributions against those backgrounds.

These, then, are the steps to follow. A. Find where Dr. X's papers are kept, which university libraries hold them, which public or private archives carry relevant information; B. Organize yourself, and spend some time in these places. Write down in commonplace books or in your laptop, each with its table of contents to be compiled as you go along, the location and the contents of each document. Copying segments in longhand, rather than just relying on photocopies, is an extremely useful if tedious exercise; C. Find what your angle is going to be: imagination tempered by a critical mind; D. Prepare an analytical table of contents. Order it in sequence, logical, chronological or otherwise; E. Flesh-out this table of contents and start writing the book.

Start writing early. Focus and angle will be modified along the way, as new material appears. Do not bother collecting and reading all the sources from the outset. Pick some parts to give you a picture of the period, or a part of your subject's life. Write parts from such material, and only then extend your searches.

The component chapters include those topics: a brief family history, predating Dr. X's birth; a short chronology, as the material for an early chapter or page; Dr. X's childhood and education, origins of Dr. X's scientific vocation, Dr. X's mentor(s) during the formative years, the various phases in Dr. X's career and a retrospective look at Dr. X's achievements.

Your subject's life will dictate the relative importance of existential details and of the scientific achievements. The Swede, Thomas Söderqvist, is a firm believer in the importance of fleshing out a scientist's biography with even unsavory details. His biography of Niels Kaj Jerne, the immunologist, describes him as an alcoholic who constantly cheated on his wife. The intimacy is startling, perhaps even shocking.

A word of caution: respect for the privacy of individuals is called for. Besides the law of libel regarding living persons, some of the information you will become privy to is best kept confidential, or alluded to without spelling out all the detail. Resist pushes toward sensationalism from for instance your publisher. Some of the juiciest bits are best omitted, you do not want Dr. X to be reduced to a caricature. For sure, your biography ought to show a bones-and-flesh person. Your subject's mind was at least as important. Dr. X may have been garrulous, a diabetic, the life of the party or promiscuous – all such personality traits are relevant. They need neither to be glossed over or overemphasized.

REFERENCE: T. Söderqvist (2003) *Science as autobiography: the troubled life of Niels Jerne.* Translated by David Mel Paul from the Danish edition, Hvilken Kamp for at Undslippe (1998) and then abridged and revised. xxviii + 359 pp. Yale University Press, New Haven, CT.

to intrigue 

BLURB

I cover here both kinds of text for the back cover of a popular book, the blurb and the endorsement. The blurb reads easily. This short text, perhaps 50 words maximum, much better yet if less than 20 words, consists of simple declarative **sentences**.

It is a most challenging piece of writing, and not only on account of its brevity. It tells what the book is about. Keep in mind though, were it to do its job to perfection, the reader might decide: »why buy this book, when I already know what's in it?« Thus you need to do a selling job as well. Explain the material while also concealing it. You want the handler of the book to be intrigued by the blurb, enough to want to buy the book.

To intrigue is a major reason to avoid mundane and trite statements, such as the present sentence exemplifies! Strive conversely for an elegant formulation of where the book will take its readers.

An actual example is Stephen Jay Gould's *Time's Arrow Time's Cycle:*

> *»In Time's Arrow/Time's Cycle Gould's subject is nothing less than geology's signal contribution to human thought – the discovery of »deep time«, a history so ancient that we can best comprehend it as metaphor. He follows the thread of metaphor through three great documents, marking the transitions in our view of earth history from thousands to billions of years: Thomas Burnet's four-volume Sacred Theory of the Earth (1680–1690), James Hutton's Theory of the Earth (1795), and Charles Lyell's three-volume Principles of Geology (1830–1833)«.*

This blurb conveys the work of a scholar, with intellectual history as his topic. It deals with geological time, so immensely long that we have difficulty grasping it intellectually. This blurb does a great job, it speaks to the self-respect of the reader.

Remember, this is a crucial piece of writing. Booksellers depend on it for their decision to order or not. This applies to both actual and virtual, online bookshops such as Amazon.com, which will be likely to reproduce your text verbatim on the Web. The laziest among your reviewers will content themselves with a paraphrase, without actually reading the book they are supposed to report on.

The endorsement deserves mention as well, whether you benefit from one or whether you are requested to supply one. If the former, you will call upon a respected public figure. When you contact this person, at the time page proofs become available, your cover letter should stress how privileged you would feel were your book to receive such a boost. Help him or her draft this crucial **paragraph** by pointing to what you feel are those special strengths of your text and which the endorser might most feel attuned to.

If the latter, you are facing something of a quandary. It is a rush job since your short paragraph is urgently needed. yet, you cannot afford to supply only bland meaningless text. You will have to read the book very carefully, to come up with a fine piece of writing, up to your reputation and consonant with the work you are endorsing.

REFERENCE: J. Davies (2001) *Communication skills.* Pearson Higher Education, UK.

include a wart 

CAPSULE BIOGRAPHY

You are often asked to supply a short **biography**, instead of your curriculum vitae in full. It is useful to have one at the ready. Not that you should send it out as it stands. Adapt it to the particular needs you face, do not supply everyone with the same standard text.

Anyway, it is a useful exercise to condense the gist of a life and career in just a few **sentences**. It calls for a talent akin to that of the caricaturist who, in a few strokes captures a resemblance. Likewise, by focussing more on the quirks of a personality, you will evoke that person.

Your text should be from 1,000 to 5,000 characters-long. Consisting of a single long **paragraph**, as a rule it should be complemented with a single portrait photograph. Since you are answering a request for a short biography, you will be given precise instructions as to the format, length and type of presentation desired.

Such capsule bios have many uses, from accompanying a proposal (research proposal, proposal of a magazine article or of a book) to providing the substance for a book cover, or an entry in an encyclopedia. You may be called upon to provide it for yourself, for a colleague (when nominating someone for an award) or someone deceased (for a study of a historical nature).

Start with place and date of birth and continue, chronologically, with the education, marriage and children if any, cause and date of death (if relevant). In-between, state the main appointments over the course of the career. Select no more than three awards or prizes worth of mention.

For a model, if one is needed, you could do worse than reading some obituaries in British newspapers such as *The Times* or *The Guardian*. They have lifted this type of writing into an art form.

Try to include a wart. Remember the punch line, at the end of the movie *Some Like It Hot*? »Nobody is perfect«. For the sake of realism and verisimilitude, your minibiography needs such a detail. Once upon a time, *mouches* were in fashion. Elegant ladies would make-up a black mark on their face, to emphasize by contrast their beauty. That's the idea.

REFERENCE: T. Schwartz (1990) *Complete guide to writing biographies.* Writers Digest Books, Cincinnati OH.

set-in-stone facts 

CHILDREN'S BOOKS

Is there any more important channel and task for science communication? To infuse young people with curiosity about the world, to suggest to them what science is about, to whet their appetite for giving themselves more of an education: those are the goals.

Let us consider each in turn. The reason for the first is an otherwise jaded curiosity. While it is true that children, as soon as they speak, raise questions continually, such natural curiosity may become stifled. Factors include adults incapable or unwilling to answer questions; poor school teachers and/or books; seeding of stereotypes by lousy television programming. This is all the more reason for rekindling such curiosity.

There are entire lines of children's books which take up actual questions posed by children, such as *Why Is the Sky Blue?*, *Why Is It Dark At Night?*, *Why Do Flies Walk Upside Down?*, or *Do Whales Have Belly Buttons?* If you take such a tack, make sure to emphasize – even though this is not easy – that the answers are not that important. Try to foster instead a spirit of joyful enquiry and a playful, »what if?« attitude. In so doing, you start fulfilling the second aim. Of course, another way to bring home actual science – as against mummified ancient science – is through **biographies** of scientists. Unfortunately, this is at present an antiquated genre. It begs for new life to somehow be breathed into it.

More generally, you will very quickly find there is a plethora of unexplored approaches for bringing science into the mindset of young readers. Others are barely touched upon. The experimental approach, strange as it may seem three centuries after the onset of experimental science, is one. A few among existing books carry hands-on sections, safe little experiments for children to perform. This is excellent. What is not wonderful, though, is that these consist of closed experiments, with recipes to be followed to the letter for a given result to obtain.

What about open-ended experiments? For sure, fear of lawsuits explains their absence. As scientists, we know how much more important it is to train the mind and the hands both for keen observation, for nurturing the unexpected, by systematically modifying the conditions and by welcoming aberrant states of the system.

The most widespread current misconception about children's books on science is likewise to view them as repositories for set-in-stone facts, definitive and irrefutable, which adults have learned and are passing on to their children.

The other wrong-headed approach to science for the young mind, as already alluded to, constrains the subjects within a few traditional areas such as natural history, zoology especially (books on big fierce animals such as dinosaurs, tigers or sharks), astronomy and space exploration. The attendant science, viz. 80% of the knowledge and 100% of the attitude (description of natural phenomena instead of accounts of laboratory experiments) is hopelessly behind.

There is a difficulty with a widespread misconception of science, stemming from the enduring attraction of natural history.

This consists in viewing science as a mere inventory of nature: no less, but no more. Mankind, in this mental attitude, is made of heirs to the natural riches. To tamper with nature, as with experimentation, is resisted. We know, though, that the world rather than just a given, is a bunch of things to be played with and altered.

The third goal listed at the outset is whetting the appetite for a science education. This is where you need to hone your storytelling abilities, starting with your own children and grandchildren. You want to implant a sense of wonder in young minds. Recapture your own sense of wonder in front of a phenomenon. Communicate the exhilaration at experiencing a suddenly fresh understanding.

Make your writing simple, uncluttered, unpretentious, invisible. Use simple words and simple sentences. Read them aloud. Each **sentence** has to be utterable in a single breath. Each page should have an astounding revelation, so that your audience will say, with jaws dropping, »no kidding«. Remain conscious at all times of this attentive and eager audience of children around you, drinking every single word of your arresting tale.

Be bold. A good criterion of having penned a good, original and worthwhile story is the inability to pigeonhole it easily into the categorization by age group (such as 4–7 or 5–10). Librarianitis is a disease, of epidemic proportion, affecting hundreds of thousands of people worldwide. They are unable to function without labels on the shelves for books. Such a condition enfeebles the patient. Of course, it is totally antinomic with science constantly entering virgin new territories. In terms of books being written, it ensures their mediocrity. Be bold, I repeat.

REFERENCES: A. Lurie (1990) *Don't tell the Grown-Ups.* Avon Books, HarperCollins, New York NY. ; P. Rubie (1996) *The elements of storytelling.* Wiley, Hoboken NJ.

secrets of success 

COLLEGE TEXTBOOK

It may change your life. If successful, since your publisher will need new, somewhat revised editions in order to sell many more copies, you may end up spending your whole time on the project.

You present a whole field of knowledge. What an assignment! This demands mastery, an overview, plus an accurate touch in rendition.

A major part of the difficulty is the »stop-time« aspect. Whereas science by nature is in constant flux, you are required to make it look static and invariant. The book gives the illusion that time has stopped and that science has reached a standstill.

Your customers are the instructors. In their majority, they are conservative. They tend to teach what they themselves had learned. Hence, science has moved away quite a bit from the material they are comfortable with.

You cannot ignore them however. To cajole them may become a need, as dictated by marketing considerations.

Ignore current fads. Concentrate on the permanent core of your discipline. Focus on the main organizing concepts. Do not become overwhelmed with facts. Make rationality and its clear expression your virtue. Show the lines of forces in the discipline. Structure the book in easy to read and relatively short chapters, 15–20 pages maximum. The first edition should be short, subsequent editions are likely to lengthen it.

Do a full mapping-out of the subject matter before even beginning. Choose very early on the chapters you will be writing first. They will serve as the sample chapters to be sent out for review.

In writing, concentrate on short, easy to understand **sentences**. Besides non-negotiable clear writing, work hard at providing good **illustrations**. Even though this will be your publisher's explicit job, you will have to help them put together the material. One of the secrets of success in this particular medium is repetition. This was a major feature in the pedagogy of the Jesuits, who were the experts in teaching the Western world for centuries.

People tend to remember better what they have heard twice. However, be astute and a little oblique in providing it. Try for it not to be too blatant.

Put at least 25% of your effort into devising the exercises at the end of the chapters. Do not plagiarize exercises from existing books, however strong the temptation. Come up with your own exercises. This is the part of the book where originality, not only pays but is a must.

Conversely, in the rest of the book, do not attempt at being original. It is a waste of time. The publisher will return you forcefully to the straight and narrow.

Too pessimistic? New departures are sometimes achievable. The wind may blow in that direction, in which case I envy you your luck.

REFERENCE: A. Haynes (2001) *Writing successful textbooks.* A & C Black, UK.

an admiring stance 

HISTORY OF SCIENCE BOOK

Quite a project! And one frought with major difficulties. Be sure that the incentive for undertaking such a project is overwhelming. Choose an innovative topic or find a fresh perspective to view it from. Select important questions to answer. Write a narrative in an engaging manner. Address the two extremely different constituencies, fellow-scholars and general readers, while giving each their due and yet creating an integrated text.

Historical writing must steer clear of two traps at polar opposites. Charybdes is the recapturing of the spirit and personalities in the given slice of time as if nothing had occurred afterwards. Such retrieval of the past, narrowed to the perspective of the participants, suffers from an obvious lack of continuity with that which followed. Scylla is whiggish history, seeing the past only in the light of the present. Such a retrospective look treats the chosen period as a mere moment in a continuous series leading to the present. Often, it is further spoiled by the illusion of progress. Typically, the former is the spontaneous style of professional historians, while the latter is how scientists tend to naively view history in their own field.

What makes science history interesting? Identifying unusual, hidden or lost roots of our present scientific culture. To stick with the metaphor of roots, if science history looks at a landscape covered with trees, it has a duty to identify the forest(s) as well. This is a rational imperative, to seek meaning in an accumulation of disparate facts. In treating your material, besides finding a periodization in consonance with it, you need to identify trends, trains of events and thoughts to structure the narrative with.

Who are your main characters? This is a good question to ask yourself. Are they the human actors? Technological changes? Beliefs and ideologies? Conceptual breakthroughs?

If human actors were indeed at the leading edge of change, carefully balance your assessment of those people in-between an admiring stance and finding each of them closeted in a niche built from a mix of prejudices, prior knowledge and education, and the then fashionable issues.

A current buzzword is »context«. We owe its renewed importance to sociologists of science, whose takeover of historical studies has modified the lines of force through the field. It is deemed essential, nowadays, to steer clear of traditional history limited to biographies of individual scientists. To bring in the social context, the material culture of the day, the strange customs of the scientific tribe – all such ingredients indeed belong with the fabric of the historical narrative.

One of the chief merits of science history is to remind us of the extreme difficulty in developing new concepts and having them accepted. What often counts is not so much the correctness of an innovative idea but its perceived utility. Such struggles are also a major resource for your narrative. Show the shock of discovery, the difficulty contemporaries had in coming to terms with it, the efforts by the author to convince peers of the validity of the find, of its importance in overturning the accepted views – all of that makes for vivid storytelling. There is also the all-too-frequent and fascinating premature discovery (Gunther Stent), at first unrecognized, and which took considerable time to sink in and to redirect science in the particular area in which it occurred.

But I am getting ahead of myself. Much work will be needed before you can start telling a story. First, gather the documentation from libraries, archives, laboratory notebooks, correspondence, science museums, collections of instruments, ...

How does one know that the research is complete? This is an arbitrary decision. However, it is of the utmost necessity lest one invests a lifetime gathering material for a book never to be written. Heed the diminishing returns in your research. When a day's worth of work does not yield a single useful piece of additional documentation, this is a sure sign that you are approaching the writing stage.

Tell the tale from the material you have amassed. Organize it from having identified the important issues, the main characters, an apt periodization and the sequence of events which made and still make sense.

In writing the chapters, aim for 15–25 printed pages as a maximum for each. Your book ought to contain no more than 15–25 chapters, hence a total of no more than, say, 300–600 pages after including the preface, **acknowledgements**, table of contents and the **bibliography** and **index**.

Having put together a historical monograph brings with it twin satisfactions. Your public reward is to have produced an authoritative work which may become the standard reference in the subject for many years. Your private reward, no doubt even more important, is the personal satisfaction of having reconnected discontinued threads.

REFERENCE: a now dated but valuable general history of science, introducing aspects of the social history too, is that by S. Mason (1962) *A history of the sciences.* P. F. Collier, (originally in 1953 entitled *Main currents of scientific thought*).

your family roots 

JOURNALIST INTERVIEW

The article, or the piece for the media, carries the name of the journalist. This is not a publication of yours. If you give the journalist respect to as a fellow-author the interview will be a success.

A successful interview is one after which you have the feeling of having made a new friend. Striking a personal interaction is a prerequisite, I find, to having a nice article written about yourself – since your work is part of yourself. One hour is more than enough time to present yourself and your work adequately. Do not waste time in idle chatter or in singing your own praises.

Do not be offended: to a journalist, science often is quite dry. Your interviewer will be looking for something to liven it up. Best of all is a joke, or something ironic or humorous. If something hilarious happened in the laboratory, or in class, then that anecdote has got a lot of chance of being the beginning of the story – which is the part that most people read.

At the very start, whether you are meeting in person or on the phone, the journalist will find out who you are. We are not talking about institutional affiliation here. Only in sheer desperation will the piece open with »Dr. X, who works at University Y …« The journalist would much rather start the piece with »Dr. X, who also collects Meissen china …«, or »Dr. X, who has just returned from Tasmania …«, or »Dr. X, the mother of red-headed twins about to enter college …« You get the idea: be prepared to supply an interesting by-line about yourself, your tastes, your hobbies. The more unusual, the better. By the same token, the journalist may become fascinated with your family roots, again if the history they carry enriches your portrait as it will appear in print. Be brief. Remember that the journalist is interested in you, not in your parents.

Remember to provide your interviewer with a calling card – not only for the e-mail address and phone number, also for how to spell your name. You should also have a handout at the ready. Do not stuff reprints of your papers into a folder or an envelope to hand to the journalist. They won't ever be read. A single sheet, typewritten on both sides, is more than enough. Any more documentation is superfluous. What kind of information should it bear? A brief professional **biography**, a dateline with the indication of your five or six most important contributions; and a one-paragraph statement of the results to be conveyed to the public.

Journalists love cuttings. So, if you have a file with old interviews and stories about you and your work, ask if the journalist would like a copy of any such material for research purposes. Journalists are insecure folks and always want to check what other journalists have written. Also cuttings are good for quotes, of the type »20 years ago the *New York Times* predicted that Professor X would win the Nobel prize …«.

During the interview proper, answer any single question directly. If you do not care to answer a particular question, just say so. No further explanation is needed. Remember that, if you announce you are talking off the record, yes, the tape recorder will be switched off. However, you will be switching-on, simultaneously, a yet higher level of attention.

Journalists have, by training, a good nose. They can smell a rat, which is why you have to be genuine. Likewise, they can smell dirt from a mile away. To them, investigative journalism is the shining path to stardom in their profession. Never mind that the scientific life usually rewards an investigative journalist with only the dullest of stories. Journalists also love academic feuds, they love arguments, disputes and fights of all kinds. Academic controversies are especially tasty since the world of academia is supposed to be so serious and civilised. Do realize that a journalist will try to turn any disagreement into a feud – for a better story.

Speaking off the record means that you may end up finding under someone else's pen the story, reflecting terribly on one of your colleagues, which you told under the cover of anonymity. You'd be surprised as to how easily the wind blows off such covers.

Do not, repeat DO NOT blow your own horn. Resist telling your interviewer that you are about to cure cancer, or solve the problems of the planet. Try to make your story interesting in itself, so that it won't need being embellished with such all-too-familiar false promises.

Journalists like short, snappy, provocative remarks. I think you should be as bold as you are happy with – maybe it's worth pre-preparing a bold remark. You can do this without sounding arrogant – you want to sound absolutely passionate about what you say. I think that's the key – you need to show some emotional love of what you are doing. Then you become easier to write about. It is very simple: it boils down to building bridges between the intellect and the emotions.

If indeed you have made a new friend, the journalist will clear the piece with you, by phone probably, on the eve of publication. Do not make any attempt at censorship, however slight. Content yourself with checking the factual accuracy.

A final point: during the interview, at no point should you attempt to steer it.

REFERENCE: B. Kovach and T. Rosenstiel (2001) *The elements of journalism: what newspeople should know and the public should expect.* Three Rivers Press, Crown, Random House, New York NY.

an impression of nonchalance

MAGAZINE ARTICLE

The public is richly ambivalent about science. It fears the very word. It assimilates science with its offspring, technology. People relish the products from technology while being all-too-conscious of its ill effects. But they are intensely curious and enthusiastic. Not only are they very willing to learn new things, they take to innovative knowledge like fish to water. It is our moral duty to answer this need on the part of our fellow-citizens.

How to go about it? From the very first paragraph of your piece, relate to your readers. You need not massage their concerns such as their health. Rather than entering that game, be imaginative. Take as your point of departure an observation from daily life. It will make a fine opening, one which people will appreciate. It will negate their expectation, of too involved, too complex and difficult a level of understanding.

Bring to the fore those qualities of yours as a scientist which a science journalist lacks. You have expertise. Bring to bear the judgment which goes with it. Do so without sounding dogmatic, an obvious risk. You carry insider's knowledge. Wear it lightly. Do it in an anecdotal manner. Demystify the persona of the scientist. Tell an episode of laboratory life in which you were a fool, before knowing any better.

Stay simple. Do not attempt to communicate more than three separate ideas in your whole piece. To do justice adequately to a single idea is already a small feat.

Distinguish between facts and their interpretation. This is a tough part of the assignment. The public is unfamiliar with the notion of theory-laden facts. Moreover, the facts to be mentioned must somehow relate to the common experience of everyone, rather than belonging to a complex experimental setup.

Another big difficulty is the dispelling of stereotypes. Your readers not only have a yearning to know more, they look at your piece with a jaundiced eye. The topic you wish to address had better be brand-new, so as not to run into opinionated prejudice. Make your facts uncontrovertible. If you attack a widespread opinion, however ludicrous in your eyes, do so with a light touch, with irony and wit.

Leave the ivory tower behind. Avoid talking down to your audience, you are addressing equals. Your readers may not have had your education, they lack your specialized knowledge. Nevertheless, they are very smart and they will find aspects in your subject matter which, by force of habit, you had never thought about. Collectively, their intellect dwarfs yours. Accordingly, show respect in all the details of your writing. Make it a submitted report from the field, not an address from a pulpit.

The title, the subtitles and the illustrations, all such ancillary material, are prerogatives of your editors. However, they will be grateful for suggestions regarding the illustrations. When you gather them, privilege the informal over the formal. A cartoon which one of your coworkers has posted will be far superior than a graph from one of your publications. You need to assemble about four times as many images as will ultimately be published – and none of them may be suitable.

Need I add that a lovely, chatty style is absolutely mandatory?

REFERENCE: K. Ackrill, ed. (1993) The Role of The Media in Science Communication, The Ciba Foundation, London.

a cameo piece 

NATURAL HISTORY VIGNETTE

The genre is a perennial favorite, it already existed during Greek and Roman antiquity. People love to learn new things about an animal or a plant, whether familiar or exotic. The genre is descriptive, a cameo piece. It has a few, flexible rules.

Your intent in such a vignette is to entertain and to instruct – in that order. To help you with the former, your text will thrive on analogies. You will compare the described organism to other things, a priori more familiar to your readership. You might have chosen to write on alligators, as Natalie Angier did recently in a wonderful vignette for the *New York Times*. She refers to them as »*souvenir dinosaurs*«, as a »*log with teeth*«, or as »*a handbag waiting to happen*«. Moreover, with respect to their formidable-looking armor, they look shielded like a tank, a panzer tank – expressions we find under Angier's pen. Were one to heed conventional wisdom, these are scary reptiles. Angier goes out of her way to dispel the prejudice. She compares them to a pet, to kittens. Their maternal behavior in looking after the young evokes a mother hen. They are bird-like in other respects, such as the anatomy of their heart.

Let yourself indulge in wordplay, too. In that same piece on alligators, Angier mentions sensory organs allowing these animals to detect even minor disturbances in their aquatic medium, which might signal the presence of prey, »current events«, as she writes, amusingly but memorably. Thus she labels alligators, tongue-in-cheek, as »*newshounds*«. Wordplay thus leads to memorable formulas, the French writer Francis Ponge in his *Parti-Pris des Choses* pioneered such felicitous descriptors.

As for the latter part of your brief, to instruct, you will astutely introduce repetitions. This was the age-old ploy followed by the Jesuits as teachers. They thus drilled the elements they deemed important into the heads of your pupils. Angier thus focuses on the large number of teeth sported by alligators (80), on their close evolutionary relationship to dinosaurs and on their proximity to birds, on the outstanding quality of their immune system.

The whole purpose of a natural history vignette is to follow the traditional format (morphology; anatomy; reproduction; nutrition; preferred habitat; classification; etc.), while at the same time dislocating and masking it. Another French writer, the naturalist Buffon in the eighteenth century, initiated such variants, in order to avoid compiling what would otherwise be an extremely tedious catalog of living forms. His *Histoire naturelle* in many volumes antedated our best-sellers.

Nowadays, a natural historical vignette would be incomplete without quotations from two or three specialists, interviewed by phone about the featured species. The relatively new discipline, ethology, i.e., the study of behavior has thus become a must for a vignette on animals. The other near-compulsory component of the piece is an image, with a caption which at the same time sums up in pithy manner the contents.

REFERENCE: N. Angier (2004) *The hidden charm of the crocodile.* International Herald Tribune, October 28.

Provide the facts 

NEWSPAPER STORY

All your wits may be needed to avoid a *lose-lose* situation. The story may not get across to the readers. Some of the elements will be omitted or, worse, grotesquely distorted.

Hence, think carefully before doing it. To succeed at transmitting a message through this particular medium is quite a challenge.

First, find out the planned length of the story, how many words are allotted to it by the editors. This piece of information may be difficult to obtain. For one thing, editors have only a rough idea in advance. They need the flexibility to publish other news which, by definition, is unpredictable. They want to deal with your story only after it is written, which will allow them to remove all the parts which they deem uninteresting to the readership. You won't have a say in such revisions.

Next and most important, have at the ready at least one telling image of your work – not just an image, an arresting one. You want readers to clip out this image, to find it significant enough that they will want to save it or mail it, and to share it with others.

The text of the story has a kindred need. Devise what newspeople term a »hook:« The first **paragraph** of the story has to, somehow, grab the interest of the readership. Try to think of a human element for this hook.

Unfortunately, you have no mastery over another two key elements which remain the prerogative of the editors, the **title** they will give to your story, and the one-liner summary, underneath the title.

How to collaborate with a journalist? It is your responsibility to make it a team built on mutual respect. He understands the needs and expectations of the readership. Conversely, you are familiar with the sources, and you can sieve out sound information from the irrelevant or untrustworthy.

Science has the potential for being turned into outstanding copy. Phenomena can be so weird, discovery is so unpredictable.

But first, you have to devise a fine storyline. What are the ingredients in a good story ? It may be human interest, which is why you should bring in the characters from the laboratory. It may involve a riddle, but one understandable to the readers when stated. More generally, your story will most likely draw on science as problem-solving. You would turn it into a kind of whodunit.

Devise an unusual plot. All the better if the actual episode you are recounting conformed to that formula. The journalist will step in to couch the story in a transparent style. That is a must.

There are three inescapable requirements to putting together a newspaper story. You need to find an angle, to offer readers a take-home lesson and to come up with a punch line at the end.

REFERENCE: I. Broughton (1981) *The art of interviewing for television, radio and film.* Tab Books, Blue Ridge Summit PA.

great enemy to creativity

PLAY

Science communication as a play is far from easy. While it can be great fun to compose a play, it takes considerable talent and work to succeed. The main reason is that the phrase, theater of ideas, is a near-oxymoron. The audience of any drama is ready to be moved by a show, to cry or laugh. However, any effort at teaching them anything will fall on deaf ears.

Accordingly, prepare to entertain before you try to educate. The former is a necessary condition for the latter.

Writing a scientific play, not only brings very little credit to its author from colleagues, it also carries a big risk of never being published. Except for highly successful plays, such as those by Bertold Brecht (ghostwritten by the women he exploited), Michael Frayn or Tom Stoppard, to mention only these names for orientation, publication of drama is not financially viable. Why? Common sense, the great enemy to creativity in general, is to blame. The common sense answer is the conflict between the supposed seriousness of science, which aims at making true statements about the world; and the supposed levity of theater, seen as only entertainment, aiming at emotions rather than at the intellect.

But you wish to proceed nevertheless. You want to try writing a play. Hence, you have already overcome such reservations. You may find them naive and mistaken. Your motivations for choosing drama as your idiom are diverse. You may wish to dramatize an abstract issue by having your characters discuss it, or you devise a play in order to plug a hole in scholarly knowledge.

A play, just like a novel, is indeed an excellent medium to fill gaps in historical knowledge. Most playwrights who have dealt with science have chosen to do so through biographical portraits of a great scientist. To give a few examples, Marie Curie, Fritz Haber, Antoine Laurent Lavoisier, Julius Robert Oppenheimer, and Alan Turing have thus featured in plays.

Historical plays, those focussed on a science episode, can thus flesh out conjectures. They allow the playwright to stage likely occurrences, to bridge the gap between the plausible and the actual. Even though the incidents and the conversations are imaginary, they are likely. They might have occurred, just as you describe. Who knows? But of course, a play is primarily a work of art. This accounts for the liberty many a playwright takes with history, adjusting it for the requirements of the dramatic action.

At or very near the beginning of your play, avoid presenting the topic, even concisely, in some sort of declarative monologue. You need to tell the public what the play is going to be about: the questions it addresses, the period it is set in, the main protagonists … To draw the audience along, you need some ploy, an enigma to whet their appetite. Withholding revelation will leave them hanging, holding their attention. Then, fill in the play proper.

For that, you need a plot, i.e., a story, characters who will enact it, their interactions, conversations, discussions, chatter, reflexions, exclamations, half-sentences and *bons mots* – anything that will instill life. Keep the plot simple. It must be semi-fictional, based on authentic facts, and still imaginary to some extent. Find the right balance.

Devise portraits of your characters considerably more elaborate than the puny fraction which you will be able to use. Only in this tip-of-the-iceberg manner will you avoid the pitfall of the cut-cardboard characterization.

The audience of a play takes pleasure in watching the characters in a wide variety of situations. But the key to success is in the dialog. Sweat over it. Have an ear for conversations. Write down what people say in your presence. Cultivate the art of the witty repartee. The dialog is a playful component, it will be the heartbeat of your play. It brings in banter and tease, it should show irrepressible wit and wordplay. Make verbal exchanges brief.

Avoid at all cost long declarations. Remember, what people say is not always predictable. Insert seemingly extraneous talk for verisimilitude.

Writing for the theater calls for a musical ear. The interaction between characters resembles that between players in a chamber music ensemble. The Haydn quartets are a prime model. Distribute between the characters an argument, a demonstration: have several voices utter it, a trio or a quartet preferably. Composing duets is trickier, they tend to become duels. They should not remind spectators of a tennis match. Polyphony provides interest. When linear and univocal, a play is inaudible and tedious.

Scenes subtly echo earlier scenes. You may enjoy structuring a play, which entails inserting such symmetries.

Go through a number of successive versions. A dozen is not unheard of. Cut ruthlessly, from one draft to the next. As the most demanding test, read your play aloud or have someone else do it. A play which does not work without all the bells and whistles, the costumes and the stage set, won't work, when accompanied by those props.

I participated in writing, jointly with Carl Djerassi, a wordplay entitled *NO* (for nitric oxide). His idea of a wordplay is great. A wordplay is meant for the classroom, for students taking a scientific course in high school or as college undergraduates. The duration of such a play should be about 50 minutes, to fit into a typical science class.

The number of characters is small, just a handful. Students volunteer to participate. They don't have to learn their part, only read it. In general, wordplays deal with social issues posed by scientific innovation, such as test-tube babies. The classroom teacher organizes the wordplay, and afterwards he or she may have the class discuss the contents.

REFERENCE: C. Djerassi and P. Laszlo (2003) *NO*, Deutscher Theaterverlag, Weinheim.

our innermost feelings 

POETRY

Although an unusual means of science communication, a poem might sometimes be the best, the only way to transmit a mix of emotion and cognition. As scientists, we try to gain a better understanding of the natural world. We report findings not only to fellow scientists, but also to a wider audience. We may want to make the general public privy to our innermost feelings, as we go about our work. The poet-scientist transliterates: this expression means putting into words what is elusive and magical about a phenomenon.

Such an undertaking is frought with near-impossibilities. Is there not deep-seated conflict between polysemy, i.e., the multiple meanings of any word in the language, and the unambiguous meaning of any scientific term? Accordingly, when communicating science, we have to recast our understanding into everyday language. It ain't easy.

It is daunting. Which is good, once turned around. We wish to convey the awe from the complexity of nature or, conversely, the elegant simplicity of physical laws. The sheer difficulty of scaling poetic expression is thus on a par with getting to grips with such an understanding, with self-understanding too, at the edge of cognition giving way to emotion.

Hence, a good poem is irreducible to full analysis. It retains a core of opacity.

Hints and recommendations? You might do worse than start by attempting to write haïkus and prose poems. Have a dictionary and *Roget's Thesaurus* at the ready to look-up words. Any poem needs considerable rewriting. When you have completed a first draft, put it away for a few days and come back to it. Your brain will have processed and reprocessed it, you may well come up with more felicitous expression of thought, with more memorable a line.

Read poetry by other writers. Use the three kinds of reading, taking in a poem whole, pen-in-hand analytical, and aloud. This is of the utmost necessity. Any poet responds to poems already written by others, even more so than coming up with a personal utterance. In a concert of voices, one gradually finds one's own, distinctive voice, its characteristic lilt and meter, and the topics most congenial to it.

REFERENCE: (2005) *Science and poetry.* Interdisciplinary Science Reviews 30(4), the whole issue of December is devoted to the topic.

concise and witty 

PRESS RELEASE

You have a result exciting enough that you want to publicize it widely. In that case, meetings with journalists require preparation of a press release. Such a document should be kept short. A single piece of paper, i.e., fewer than 3,000 characters if printed on both sides, ought to do. Remember you are writing for the uninformed, not your peers. Accordingly, your results have to be not only rephrased, they have to be reframed. They have to be put into perspective, one that the journalists will be able to grasp readily.

Choose an angle. The possibilities are endless: human interest; animal life; difficulties encountered during the research; the teamwork having made it possible; the improved understanding thus achieved; pluridisciplinarity; ethics; etc.

A pitfall is using this pulpit to denounce the lack of adequate funding you suffer from. This won't wash, you are attempting to instrumentalize the journalists for your own ends. It is not the proper medium for complaining about the material needs of your laboratory. You want to contact people in your granting agency for this purpose and address the issue with them.

In writing the text which will present your exciting new results, flesh it out with what will turn it from an ordinary piece of advertising – let's face it – to an interesting document. Only you can pull this off. Was there a serendipitous element in your success? In that case, refer to it in a concise and witty way. Does one of your coworkers deserve most of the credit? Give the name and tell very briefly of the circumstances. Did you prove other people wrong? Do not gloat about it. You get the point: an approach, modest, terse, matter-of-fact, is the only one capable, not of making you conquer the media, but of giving you a passing grade, of ensuring that you meet at least some journalists.

Each journalist who answers receipt of your press release ought to be met in person. Try to set up such a meeting. Otherwise, arrange for an interview on the phone or by e-mail.

Hence, the press release should end with the mailing address of the laboratory, phone and fax numbers, and an e-mail address. Include a short **biography** of the main investigator (one paragraph). Include also, this is mandatory, at least a couple of relevant pictures. They ought to be both self-explanatory and visually interesting. I know, this isn't easy. Also, and this will determine the range of diffusion of your story and therefore has to be prepared with the utmost care, give names and addresses of five contacts who can and are willing to explain the significance of the advance. These obviously have to be colleagues, people who are enough in the know to validate the work, whether your competitors or not. Remember: journalists like historians are trained to check their sources.

REFERENCE: P. Austin and B. Austin (2004) *Getting free publicity.* How To Books, Oxford, UK.

snozzcumber 

PRIMARY SCHOOL TEXTBOOK

Any writing conveys a philosophy of life. As unlikely as it may seem, this will be your most important contribution in writing a text for young children at school, initiating them to some aspect of science. I shall address here the content, exclusively. Your publishers will produce a lavish **book**, with lovely **illustrations**, a nice handy format and easily legible type; so that each item, presumably an experiment to be performed by the children, as opposed to a demonstration by the teacher, receives an attractive presentation. This much can safely be assumed.

Steer clear of space exploration and dinosaurs, as topics. This ought to be an initiation to science, not a regurgitation of stereotypes. Which does not mean that you won't make a claim on the imagination of the little users of your book. To the contrary, it will be your best resource. Infuse your text with fantasy, with wordplay if you feel like it – you can do worse than use, say Roald Dahl, as your model; he coined memorable idioms such as »*swishfiggler*«, »*snozzcumber*«, »*wangdoodles*«, »*fizzcocklers*«, or »*whoopsy-splunkers*«.

Use the proposed experiments to teach the importance of careful observation, of note-taking, of making conjectures and of coming-up with ways in which to put them to the test.

Strike a balance between natural history and experimental science. Both are necessary, in about equal amounts. The former is needed to instill a sense of wonder: the maggot turning into a winged insect, beans on a bedding of cotton wool sprouting a plant embryo, ... The latter is needed to show, not only the penetration of a scientific outlook, also how much it goes against common sense. A typical experiment for that purpose is floating pins or needles on a glass of water, by way of an auxiliary such as a very thin piece of paper to rest the metallic object horizontally on the liquid.

Try to find experiments which will not only make a point, but be fun. With magnets, for instance. Or optical illusions. Children are fond of showing their newly-acquired knowledge or skills to their parents. Hence, experiments which require only a piece of string or two are to be privileged. For instance, attracting small pieces of paper with the tip of a pen or a comb after rubbing it over a garment. Or, showing how objects of various densities find their depth in a column of three nonmixing liquids, such as water (colored with a little ink or blue dye), oil and syrup (caramel or maple).

Yes, you will be hitting against a big limitation. Most of the experiments in the book won't go beyond eighteenth-century science. But this does not matter at all, what you want to impart is scientific curiosity.

The teacher needs to find in your book the seeds of ideas for hands-on experiments. Write your book, though, with the pupil not the teacher as your intended recipient.

REFERENCE: A. Haynes (2001) *Writing successful textbooks.* A & C Black, Bloomsbury, London UK.

laboratory lore 

PUBLIC READING

Just like other writers, science writers are occasionally called upon to perform in this manner. The three aspects to be mainly concerned with are the sophistication of the audience, the choice of pieces to be read and the reading proper.

Do not assume any background in science. The central paradox of science communication is that, if there is huge appetite by the public, retention is nil. Accordingly, you will need to supply the basics as you go along. Which is just fine. It will make for small asides in a conversational, an intimate tone, to enliven your performance.

Do your homework. Prepare your program carefully. If the session is to last about an hour, do not read for more than a total of 40 minutes. The discussion will take the remainder of the time slot. Hence, you should choose a variety of texts as regards length. Perhaps three or four short pieces, a single longish one, and three or four of medium length. However, more important than duration of the individual reads, is their forming a harmonious whole. Yet more important is that these pieces call upon your whole registry as a reader. A public reading can easily turn into a monumental bore. Unless some of the segments are done in a lyrical voice, others as an oratorical and yet others in a neutral and factual tone. You need diversity to prevent sleep setting in. Needless to say, but a useful reminder, you will be **telling stories**.

And forgive me for repeating this particular recommendation. Put your book aside and digress about what did not get into the published story, and why, about small anecdotes in the making of the **book**, laboratory lore, and so on. Your public is likely to be interested by small asides, and will be all the more ready to become captivated by the reading. In the same way one makes a sandwich, a public reading needs both the bread and butter, and the stuff in-between. Your performance will be highly successful if all those ingredients are present and if none steals the show.

Prior to being introduced and started, run (or walk) around the block! This recommendation addresses the need for full oxygenation of the brain and proper ventilation of the lungs. It will also calm you down to go outside, by yourself, to walk a few paces and inhale deeply.

Dress both comfortably and conservatively. Sit comfortably, which means trying out your chair in advance and having the organizers replace it if necessary. Make sure to have at the ready, a box of Kleenex, a supply of water to drink and throat lozenges. If you need reading glasses, all the better. They will supply a most useful prop, when you take them off, when you hold them while making a point and looking straight at a member of the audience.

Pace yourself very carefully. This is best done when rehearsing your talk. You will inevitably find yourself reading way too fast. Slow down. Find a way in which you will be speaking more slowly, more deliberately, more ar-ti-cu-la-te-ly and therefore much more understandably than is your normal guise.

Mark your text with a pencil: this is part of the homework. When you prepare, indicate on the page where breathing, emphasis or, conversely, a quieter tone (to convey intimacy) are needed. During delivery, if at any time there is heckling, voluntary or involuntary – people coughing, people engaging in chats with a neighbor – stop in your tracks, give a hard look at the heckler (of the »if looks could kill« kind), and wait for the person to stop being a nuisance. Resume your reading where you left it, at identical rhythm and sound level.

REFERENCE: S. W. Hyde (1995) *Television and radio announcing*, 7th edn, Houghton-Mifflin, Boston.

Make connections 

RADIO INTERVIEW

Why do it? If you don't care to do it, whatever the reason, it is far better for you to turn down the invitation. Conversely, if you feel confident, if it gives you an opportunity to express yourself on a topic of your expertise, do it.

Whether the interview is broadcast live or not has little importance. If canned, the **editing** will give it polish. You will sound even better.

The quality of the broadcast will reflect that of your homework. to get the feel of it, listen to earlier shows in this particular program. Anticipate the questions which will be put to you. Prepare in writing interesting and articulate answers. Practice saying them aloud, until half-memorized. Besides the likely questions, consider, too, those things you are keen on saying. Know exactly what they consist of. Limit yourself to three or four such points. Practice saying them as well.

Generally speaking, you will be contacted by phone for a pre-interview. This is an opportunity for you and the reporter to strike an acquaintance. At the moment of introduction, write down phonetically the name of the journalist for future reference. Both your times are precious. Use this occasion to chart out jointly the territory of the interview. Establish your expertise by the depth and scope of your statements, never in self-promotion. Use the pre-interview to educate the journalist in the subject area. This will pay off, you will be asked good questions in the interview proper.

You are now about to get on the air. You are in a studio or at your desk, if the interview is carried out on the phone. In either case, sit in a comfortable and quiet, not creaky chair. Place in front of you a piece of paper and a pen to jot down a few quick notes, if need be (phoned-in questions, for instance). Place also in front of you your list of questions and the answers. Refer to it regularly during the interview. Keep a supply of water nearby together with throat lozenges, to be used during intervals or silently.

You are on the air. Your host tells who you are, and why you are called upon. Answer briefly and await the first question. Be very relaxed. Remain attentive. Fluid speech is called for, which implies great self-control. Make it conversational, even chatty. The best interviews are run between two professionals, each at ease, who work together to create an interesting and informative show. Do not ruffle papers or bang your knees (one of my unconscious traits!) during the interview.

Make connections, this is important. Emphasize links to everyday life, to what is familiar to your listeners, to other fields to. Tell your listeners how you have acquired this particular piece of knowledge: this is likely to be the most fascinating part of the interview to them. Do not make empty promises (a cancer cure by the end of the decade) nor extravagant claims. Remain modest. Remember: whatever your state of mind during the interview, worried, angry, jocular, stressed or tired, it will show.

The interview is over. Do not forget to pack your stuff before leaving the studio. Make sure to shake hands with the technician in attendance: this person is your best evaluator, from extensive experience. He will tell you how well you did. If you managed to grab his attention, the interview will have been a stellar success.

If indeed it was, this means that your preparation was thorough. This ensures that your host will keep you in mind for future shows.

REFERENCE: R. Lomax (1982) *Writing for TV and Radio* (The Writing School Guide Series). Leisure Study Group, London.

the conjecturing 

SCIENCE NEWS

Science news is a victimized genre. Billions of people depend on it as their only source of information about science. The near-monopoly by journalists has too often produced a travesty of both the science and the news. True, it is a challenge to convey a piece of research in the form of bites, whether in print, on radio or on television. Can we improve on it? How can we scientists influence and make use of this medium to educate the public on what science is about.

To criticize journalists is the all-too-easy part. Yes, they look for the most sensational elements. They tell the public only what it wants to hear: because of the particular piece of work being reported our well-being will improve, cancer will be cured and people will live to be 120. In their effort at simplification, they oversimplify and they caricature. They deem themselves essential, as go-betweens. Indeed they will remain essential in their function as mediators, unless we show ourselves capable of an end-run.

All such reservations are well-founded. But can we do any better? For starters, can we help science journalists do a better job, if unwilling to dirty our hands and cooperate with them? This ought to be the starting point, learning to prepare a **press release** that is informative, to the point and readable. Does it rule out including all the uncertainty, the tentativeness, the conjecturing and the groping for the truth which are all parts and parcel of our activity? Can we find ways to communicate such important aspects? Or are they out of reach?

Is that so? Can we not communicate our excitement at doing science? Are we unable to turn into narrative the scientific adventure, to recount the hurdles we have had to overcome, whether factual or conceptual, and to present a story all the more alluring for being real? The context, what journalists do not want to hear and what they leave out, is our strength. The journey, not the arrival, matters. We are the only ones to know it at first hand and therefore to be able to tell it as we have experienced it. Ask yourself: why do I want to make it a piece of news? What's important to me about it? Why should the general public care? Shall I spare the time to reconstruct the story as a brief popularization, which won't be a straightforward effort and will require all my talent as a storyteller? How can I share my hands-on experience?

To educate the public means etymologically to lead it. How can we exercise leadership by means of science communication? This is close to being a moral imperative.

If the genre of science news is sick, what it will take for it to regain its strength and its credibility is a systematic attempt on the part of scientists to tell their own story, with or without intermediates. It is not that difficult, we can use as a model the sports pages of newspapers and magazines.

Communicate your enthusiasm, share the thrill of doing science, of understanding the natural world.

REFERENCE: T. A. Rees Cheney, *Writing creative nonfiction: fiction techniques for crafting great nonfiction.* Ten Spee Press, Berkley CA.

from the bottom up 

SCIENCE POPULARIZATION BOOK

Everything is in the intent. A »me-too« imitation won't fly. A **book** needs three ingredients. An author's need to get it out of his system, striking a chord with the readership, and the book taking-on a life of its own. If isolation becomes the writer, spread and diffusion are the essence of a book.

Aim for the best. Some classics of science were never written for a scientific audience but for a general readership. Darwin's *Origin of Species* (1859) is a prime example. It shows, just as with other classic books – Benoît Mandelbrot's *Fractals* for instance – that bypassing the experts and going to the general public may be a winning strategy. By »aim for the best«, what I mean is not only to be ambitious, but also to translate a masterful overview of a subject, or field, into a book which any educated person is bound to enjoy and which will have enduring value, such as for instance, George Polya's *How to Solve It* or Hermann Weyl's *Symmetry*.

Let us start with the subject matter. Choose both topic and an original viewpoint (or approach). Regarding the former, most books nowadays, 80% perhaps, are editor's books. They are written from the bottom up. The editor makes an educated guess as to what in his or her opinion, the public needs (and will buy).

Let me give the example of clays. Their chemistry is one of my areas of competence. An editor's book, and I have to exaggerate and caricature to make my point, might deal with *Getting Mired in Clay and Climbing Out, An Existential Self-Help Manual*, or *Handling Clay, the New Management Method*, or *The Claypot Cookbook.* You get the point. Too many books today aim at filling a blank. They pigeonhole themselves into trendiness, which makes them ephemereal.

Conversely, an author's book is written from the top down. It stems from a flash of intuition. On this same topic of clays, I would perhaps try to write a *From Clays to Plastic Bags*. Chapter 1 might describe child play with mud and plasticine. Chapter 2 would cover prehistoric pottery. Chapter 3 would survey plasticity of natural forms, such as an egg in the duct of a hen or a child being born, or yet a glob of resin exuded from a pinetree. Chapter 4 might be devoted to sculpture from moulding clay or plaster, as opposed to chipping away at wood or stone. And so on. The topic chosen for a book demands the three qualities of virginity, of being at the forefront of intellectual exploration, and of charting new territory for itself.

If a book is to succeed, the approach has to be original, as against what might be termed a journalistic attitude. A journalist, faced with the task of writing a book on clays and plasticity, for instance, should compile a **bibliography** of important relevant papers. Digest the collection, and summarize the content of ten or twelve papers per chapter, inserting interviews with some of their authors, reporting diverse opinions, as well as fleshing-out their personalities with portraits and the odd detail.

Only *you* are able to determine how to go about it. Had I to write this book on clays and plasticity, I might choose a crazy angle: how does it feel to be plastic, from the standpoint of the clay?

I might start with the clay platelets sitting stacked on top of one another, lubricated by water in the interlayer spaces, and thus able to slide past one another from the application of only a weak force. I would relate the ensuing plasticity with disasters of dams, inadvertently built on layers of clay giving way.

Alternately, I might pick the topic of geophagy: to this day, Americans in states such as Georgia or Kentucky literally eat dirt. The explanation is a potassium deficiency and adsorbance of toxins, a phenomenon called »geophagy« that also exists in birds and chimps … In-between lies a potential book, which, if well written, would be a fascinating read.

What types of books on science, for the general public exist? There are monographs, collections of essays, biographies of famous scientists, narratives of discoveries. In short, the field is limited to just a few categories.

The purpose of such books is to make concepts intelligible (example: relativity theory), to tell the history of a field (example: »What Does Matter Consist Of? From Lavoisier's Discovery of Oxygen to the Periodic Table«), or, much superior in my opinion, to convey scientific life from the inside, as it were (»Groping in the Dark: the Search for Vitamin C, from Lind to Szent-Györgyi«).

After having signed a contract with a publisher, one sets about writing the manuscript. There are half-a-dozen components. Deciding upon a title is the prerogative of the publisher, but you will have input into the final decision. A good title is memorable, brief and to the point. Examples include Mark Kurlansky's *Cod*, Stephen Hawking's *A Short History of Time*, Peter Atkins's *Molecules*, Sebastian Junger's *A Perfect Storm*, and Stephen Jay Gould's *The Panda's Thumb.*

Addition of a subtitle is currently in vogue. While the title is meant to be catchy, the subtitle is more informative, indicating to booksellers the category in which a book belongs.

Provide, after having written the entire manuscript, a preface. Its purpose is to spell out your reasons for writing the book and to summarize its contents. Present your approach and justify the structure chosen in organizing the material. These few pages, half-a-dozen at most, will give the reader a foretaste of what's to come. It is the place for a personal note on what this book means to you.

What follows is to be organized into sections, each part subdivided into chapters. The chapter is a basic unit, and, ideally each chapter ought to be readable in a sitting. It should correspond to one of the successive episodes in your story. Endow it with forward momentum, and at the end of the chapter put in a short paragraph serving as a transition to the next chapter.

In writing the manuscript, you will repeatedly come against the problem of technical terms. While they are to be avoided, a few nevertheless will be needed. Cull them as you go along to be gathered for a glossary at the end of the book.

It is usually a good idea to insert notes, whether footnotes or endnotes in order to answer the needs of some of your readers for scholarly references. Reviewers may be scathing if such material is absent. If indeed you opt for inclusion, it will be scrutinized. Make sure that each reference is 100% accurate. In particular, carefully check the spelling of family names. I have always admired the British tradition of the bibliographic essay at the end of a book. Instead of merely providing references, the author presents his personal evaluation of the bibliography while briefly summarizing the kind of information which any given reference provides.

Another segment of the book which needs to be written at the very end is the index. Like the scholarly appartus, reviewers will also take a close look at it. There is an aspect of the *Who's Who* in an index where names of people are concerned. Make sure that all the important contributors to the field are referred to by name in the index – otherwise you may be in for some trouble. More people than you think upon leafing through a book, a review copy in particular, will first look up their name in the index: human vanity.

Personally, I do not rely on professional indexers. You are the best person to put together your index. This is not fun, but it is worth investing the time – since a well-prepared index is the best portal into the detailed contents of your book, superior even to an analytical table of contents. The assets of an index are its low-cost as a searching device, the browsing it allows, and the quick retrieval.

The back cover, like the title, is the prerogative of the publisher for the same reason, advertising the contents of the book. You will be consulted and your thoughtful suggestions will be heeded. However, marketing and sales people will have their say too. It consists of a summary of the book in a few sentences, complemented by a capsule biography of the author, and often enthusiastic comments from some authoritative people in the field.

If you are writing your very first book, the workload to expect is divided about 50% documentation, 30% writing proper, and 20% editorial interaction (the learning part of the exercise). As a ballpark figure, expect to devote about 50 hours for producing 15,000 characters, i.e., ca. 3,000 words, or six days of uninterrupted writing.

The analogy to a child is apt. A book takes on a life of its own and requires being looked after. There may be a need for revisions with successive printings. It may be turned into a paperback, which will also entail a few changes. There may be translations in which case you would be well advised to help the translators in their task and answer their questions. Your book will be purchased by libraries, which will also contribute its share of queries from readers.

REFERENCE: W. Zinsser (1998) *On writing well: the classic guide to writing nonfiction.* Harper, New York NY.

rush to push 

SCIENTIFIC EXHIBITION

It shows typically a mix of scientific instruments, documents from archives and, often, art pieces. The location, as a rule, is a science museum, such as, to name a few, Cité des Sciences et de l'Industrie in Paris, the Wellcome Foundation in London, the Science Museum in Oxford, the Chemical Heritage Foundation in Philadelphia, or Discovery Place in Charlotte, North Carolina. It is unfortunate that such shows are held only in such precincts. It gives the wrong impression, of science as being the property of scientists, whereas of course it belongs to everyone. Accordingly, an effort has to be made to present displays in many other and quite different locations.

At present, the audience consists predominantly of families with children and of groups of schoolchildren. The latter, typically, are accompanied by a science teacher. They are encouraged to take notes. They are told to stand still while a person from the science museum gives them explanations.

Their spontaneous behavior differs considerably. They rush to push buttons more or less at random on the displays, in order to elicit spectacular responses. It is a shame that the thought and hard work that went into preparing the exhibition is totally lost on them.

In the same way that we ought to find more diverse locations for science exhibits and fairs, we need to diversify their audience. There are lessons to be learned from the successful effort at public relations by art museums, worldwide, during the last half-century.

Your most important task, if you are going to put up such an event, is to devise a theme. A sideways approach often works just fine. For instance, an exhibition about math – a scary subject – takes the approach summarized in its title »Flip It, Fold It, Figure It Out!« To quote from its description, »*Visitors can play with quilt and tile patterns, create drum beat rhythms and guess at three-dimensional mystery shadow shapes* ...« Another show, also put up by Discovery Place, takes grossology as its topic. What is »grossology?« It comes from the title of a book for children by Sylvia Branzei. The exhibition deals with human physiology. Its appeal is described thus: »*it explores all the slimy, mushy, oozy, scaly, stinky and gross (yet, scientific) things that happen in the human body every day*«. You get the idea.

Once you have devised a catchy theme, you will need to identify, contact and convince sponsors. For this purpose, you need to put together a brochure with a descriptive text (at most, half-a-page) stating the purpose of the show, a list of the exhibits, a floor map, a budget, a calendar including supplementary activities (such as lectures by distinguished guests), and a conservative estimate of the number of visitors.

Your exhibits are not unlike images in a book or magazine. They should carry their own message. The accompanying explanatory labels cannot be too brief. Aim at a maximum of 75 characters for each, i.e., about the length of this **sentence**. You can provide more detailed explanations in the catalog.

This ought to be an illustrated book, beautifully designed, with lasting value from the quality of writing and from the art, which buyers will not only display on their coffee tables but will also use at first for bedside reading, later on for reference. Many a scientist's vocation was awakened by such a catalog, an excellent reason for turning it into a masterpiece.

REFERENCES: http://www.discoveryplace.or/discoveryhalls.asp, R. Gardner and E. Kemer (1993) *Making and using scientific models*, Frankin Watts, New York.

the appeal of science 

SECONDARY SCHOOL TEXTBOOK

This form of science communication is beset with obstacles. To start with, one addresses two very distinct constituencies. Children are the users while teachers are responsible for selection. Another difficulty is that, at about the turn of the twentieth century, the educational system gave up trying to follow scientific progress. As a consequence, most of the science taught today dates back to the eighteenth, not even the nineteenth century.

Yet the task is of overwhelming importance. This is the age group at which the appeal of science is at its highest and can be stifled by poor teaching, even killed. You may yearn to try your hand at writing an up-to-date textbook. You may have an original approach. In that case, you should go ahead. Ignore however other considerations, such as it being a lucrative venture. You will succeed only on the strength of your enthusiasm and with an extremely professional job.

Assume then, that you have received a commission from a publisher through an acquisitions editor. You will be assigned to a production editor, and your interaction with that person is critical. Treat it with care. This is a necessary condition for the venture to succeed.

You will find yourself in a quandary, using new material while at the same time fitting within a rigid curriculum. If you are unwilling to accept such a strict constraint, do not even attempt writing a textbook.

Organize your writing by preparing an analytical table of contents. Make sure that it is logical, self-consistent and comprehensive. Once your material has been structured in this manner, and only then, will you be able to start writing.

At this point, line up a list of reviewers who will agree to take a critical look at your text, perhaps even for free if you turn to friends and friends of friends. Submit to them your tentative table of contents, they will give you good advice.

Your editor will likewise seek expert opinions to make you benefit from them. At the same time, your editor will be under pressure from the marketing department. Do not let such commercial considerations overrule your gut feeling as to what your book should be. In other terms, you need to think of sound commercial reasons for your decisions in order to justify them in addition to the science and the pedagogy.

With each page, the easy route is tempting, to insert a point that will make you popular with the students, their teachers or, who knows, the sales representatives. Resist such self-defeating urges.

Do not however use a highbrow mode. Make sure for instance, to include in each chapter at least one example explaining a nonobvious phenomenon from everyday life.

Your textbook should not be only a compendium of scientific results. It should also inspire a scientific way of thinking. As such, include tools for thought such as reading a double-entry table; understanding a graph; arguments from orders of magnitude; and so on.

Work hand-in-hand with your illustrator. Graphic design is just as important as the quality of your prose.

Work backwards. For instance, the exercises at the end of the chapters are, perhaps, the most important part of the **book**. Why not start each chapter by putting together questions and problems? Having done so will help determine what notions to introduce, explain and justify in the body of the chapter.

Secondary school students are both an endearing and an infuriating lot. But you can sway them. They need to know, not so much the established knowledge, but what science consists of: a groping for the truth, often counterintuitive, haphazard, a human enterprise, a constant controversy. Find ways in which your own enthusiasm will first reach them, and secondly inspire them.

REFERENCE: A. Haynes (2001) *Writing successful textbooks.* A & C Black, Bloomsbury, London UK.

witty and brief answers 

TV SHOW

A misconception common among scientists is to consider a TV show as a filmed discussion or **seminar**. This is wide of the mark. A TV show is a performance art. Unless you are a born performance artist, steer clear. You are all too likely to feel either unequal to the task or cheated. And the audience is likely to get comforted in its belief that scientists decidedly are dull people, unable to make crisp or provocative statements.

An exciting invitation, it provides you with a huge audience. But this is an expensive medium. The cost is of the order of a thousand dollars US per minute. You won't be given much time to broadcast your message.

TV shows can assume a variety of format: live or canned, with feedback from phoned-in remarks or questions, one-on-one or as part of a panel, ... the audience is eavesdropping. Members of the public need to watch effortlessly. Failing that, they will zap you out of their existence.

If you are unable to beg out of the commitment, you'd better prepare for it. It is not such hard work, provided that you switch into a rather unusual mental frame – yet the time investment is awesome. You will need 15–20 cue cards to remind you of the facts you may need to cite at the drop of a hat. You will also need – this takes ample preparation – a dozen one-liners to offer at choice moments.

As part of your preparation, rehearse at length and several times with a mentor, who will come up with outrageous comments, with questions to take you aback and to force you into witty and brief answers, rather than lengthy classroom explanations. If the show lasts about half-an-hour, you will need at least six hours of such conditioning.

Prepare what you want to say, make sure, however, to connect it to what people already know. Preparation thus needs to be extensive. A good rule of thumb is to invest about a factor twenty the air time you are programmed for: an hour of careful thinking for a three minute bit, which might be part of the evening news; several hours of reflexion and writing for a longer interview of, say, 20 minutes.

You will have prepared yourself for the likely questions. You will have groomed yourself for the most unexpected questions. You will have the opportunity to tell about three factoids, with which to illustrate results, and to impress your audience. Hence, make sure that you have double that number on the tip of your tongue, ready to quote when your host presents an opportunity. The message should be those three factoids – plus yourself (which I shall come back to).

If the budget allows, the producers may insert a documentary portion shot in your lab. This would be terrific. Yet it will be a little burdensome in terms of time lost, of demands placed on you and your coworkers by the advance team and the production people. However, it is well worth it. It gives you the opportunity to show real laboratory life to the public.

Remember Richard Feynman and his request for a glass of iced water, when he demonstrated to the public the dramatic loss of flexibility of O-rings for the space shuttle, after its disaster at launch? If you are able to pack a small demonstrative device, in your pockets or in a small package which you will unobtrusively hide by your side, bring it out at the appropriate time. This, if you have thought of it carefully, may turn into a magic moment for the viewers.

As part of your preparation, do not fail to include a seemingly improvised bit which you will have memorized, together with a brilliant final statement.

When on the air, do not anticipate any time to develop an explanation, to explain conditions of validity or how a particular conclusion was reached. Television, of course, is a popular medium. Moreover, it tells people more or less what the announcers and the producers think they want to hear. Its down side is regurgitating what is already common knowledge. Don't get caught in such a trap.

It takes two (or more) to play. Most likely, you will interact one-on-one with a professional journalist. Do not even try to outshine this professional. To the contrary, this person will do the utmost to bring out your best. Do not interfere, go along. During the preparatory phase, feed a maximum of two or three well-chosen documents to this TV journalist: non-technical, brief and to the point. The same qualities are required during the interviews you will have, on the phone or in person prior to the interview to be aired.

Your host will have put in an impressive amount of homework. Before you are on the air make sure, in words and **body language**, to communicate your empathy to this person. During the show, make the journalist sound good (»what a perceptive question«).

If live, as opposed to pre-recorded, there should be no difference in your attitude. When the journalist and the cameras turn to you, there is no reason you should feel on the spot. Take your time. Take time to breathe before you answer. Take a drink if you feel like it. Compose yourself into being at your most natural. Be relaxed, natural, speak slowly and be aware of your body language. It should exude a sense of ease and cordiality.

Dress smartly for television and a bit conservatively. Dress as if interviewing for a job. Wear a dark suit and a bright tie or scarf. Avoid a dress shirt or a blouse with stripes. Look straight at the camera, not at your notes or cue cards. Be at ease and show it. Feel free to laugh. Do not wear a forced smile. Do not hesitate to smile at an interesting or funny remark.

Indeed, viewers are focussed on you, and less on what you will tell. They are interested in your life experiences, what led you to the particular experiments you are relating, in the personal satisfaction you derive from your work, in the relationships with your associates (banter included) – in a word, they are eager to get to know you as a person. Do not worry, the journalist is aware of that need and will do the best to satisfy this particular hunger on the part of the public. When about to speak, take a deep breath. At all times, sit comfortably and behave as in conversation.

REFERENCE: I. Broughton (1981) *The art of interviewing for Television, Radio and Film.* Tab Books, Blue Ridge Summit PA.

providing a broader picture 

WEB SITE

It is a necessity. Why? It gives you the means of sharing information in a manner you control, with people you might otherwise not reach – total strangers who nevertheless may provide you with rewarding contacts.

There is a general philosophy as regards a Web site about a scientist. It should not only present but complement your lines of research. Provide a wider cultural context for each of your scientific endeavors. This will take considerable time and work. It is worth it. Providing a broader picture makes your work, and thereby yourself, considerably more interesting.

The cover page needs to present a short biographical summary, together with a photographic portrait. More generally, strive to upload lots of pictures – and be prepared for them to be lifted for all sorts of purposes, legitimate (use in classroom) or not so legitimate (which boggles the imagination).

Follow-up with a list of the pieces you are making available on the Web site. Organize the texts and accompanying **illustrations** according to a few categories such as, representative research publications, **review articles**, magazine pieces, resources for the science teacher, and so on.

Also include your schedule for the forthcoming months and any additional information you care to advertise, such as the areas in which you are willing to undertake consulting assignments.

Insert **abstracts** of popular **lectures** among the posted items. Whet the appetite of those people visiting your site (do not forget to put in a counter for the number of visitors) with factoids and Trivial Pursuit-like questions, directed to your area of expertise.

Do not forget the entertaining part. For instance, insert little vignettes, such as anecdotes about life in the lab.

Finally, don't build this homepage by yourself. Entrust the design to a professional who will bring to bear training as a graphic designer, together with skills as a programmer.

Try to update your site regularly, a bimonthly basis would be ideal.

REFERENCE: P. v. Dijck (2003) *Information architecture for designers.* RotoVision, Hove UK.

PART III

DECISION MAKERS

GENRES

BRIEFING

A briefing is addressed to a decision-maker who does not understand the underlying science, but who needs advice based on sound science. There is no significant difference between an administrator, a politician or a businessman. They all need background information upon which to base a decision. Not just an opinion, they need hard data to fall back on in case the media question them, or they have to justify themselves to the public.

The interview with this person is likely to be short. At the end, leave behind a written text – near-identical to your oral presentation. What you tell the decision-maker in discussion without a paper trace is all too likely to evaporate.

The first difficulty is to tackle the provisional status of scientific evidence. In all likelihood, the issue confronting you does not have an unambiguous solution, based on the available science. Thus, you are setting down your personal judgment. It may conflict with that of other experts. The ensuing report thus resembles in essence, the discussion part of a scientific paper. It is an exercise in interpretation.

Therefore, you need to back up your conclusions with the combination of hard data and sound logical argument. This scaffolding however is for yourself and your archiving. It has no place in your report to the authorities.

In style, make your memo similar to the abstract of a publication in a journal. As the term »briefing« says, keep it short – no more than a couple of pages overall. Write it using only short **sentences**. Insert one or two pertinent formulas to sum up the issue or your argument, or yet another position which you reject. End this text with several numbered recommendations.

Make your assumptions extremely explicit. Likewise, make any conclusions you draw depend upon these assumptions. This is perhaps the most difficult part of your assignment. Do not hesitate to prepare two or three scenarios, each consistent with the available data but leading to different conclusions. You do not absolutely need to opt for one of those scenarios as your preference. If you do, make sure to state your reasons for doing so.

If an issue or a topic is controversial, say so. State whether your views are mainstream or marginal. If the latter, be at your most eloquent and persuasive.

Do not avoid any unpleasant point, it will come up ultimately and you will regret not having spelled it out in the first place. And remember: the quality of your briefing may affect some of us during our lives.

Any dialog between a political decision-maker and a scientist is tricky at best. It is difficult to set-up and turn into a success. Sources of mutual misunderstanding are, on the part of the scientist, the inability to sympathize with someone who is not readily swayed by rational argument alone, and someone who may not even try to understand the argument you are presenting. And, on the part of the politician, lack of genuine esteem comes from the apparent incapacity of scientists to agree between themselves on anything. The decision-maker cannot understand that science does not operate through consensus, that controversy is its healthy daily fare and that, accordingly, one can always find a scientist to challenge any given finding or assertion. Most likely, the political representative shares the general opinion: there is a technical fix for everything. S/he does not understand there may be firm physical limits making it impossible or unwieldy. Neither does s/he grasp the difference between science and technology. It is better to be well aware of all such differences.

The decision-maker expects a handout from you. It ought to contain a backup report, which will carry the factual evidence. Thus, you ought to prepare very carefully, while remaining aware that in all probability it will never be read nor consulted. There should also be, on the cover or first page of the handout, a one-**paragraph** statement of what you are proposing. Make it precise. The writing has to be neutral, which does not rule out inclusion of a memorable formula.

Then there is the briefing proper. This document ought to contain, first a table of contents, referring to the numbered pages. It should consist of an outline of the argument from both sides, pro and con. This is not a perfunctory reference to the position of your opponents. It has to be honest, neutrally-written, presenting both sides of the coin.

The briefing will end with a cost analysis of the budgetary consequences of each of the positions or scenarios your document defines and fleshes out.

In your oral defense, when you present your case to the political decision-maker, prepare to answer the key question: a convincing explanation of why you are taking this particular stand and why you are lobbying for it. If you stand to gain a personal advantage of any kind, be forthcoming about that aspect too.

To return to your position paper, you will have inserted the subtlest of hints as to the advantages this individual might draw from drafting a bill which would be consonant with your position, with the feelings of his colleagues and with those of his constituency. Now, I wish you good luck.

REFERENCE: S.C. Witt (1990) *Briefbook: biotechnology, microbes and the environment.* Center for Science Information, San Francisco.

various costumes 

CONFERENCE REPORT

This kind of a text occurs in various costumes. I'll address three of them. One is the administrative report, required formally by the office which gave you funds for attending the conference, such as an airline ticket, reimbursement of your registration fee and local expenses. In that case, heed the Golden Rule regarding administrative reports: treat administrators as fellow-humans, not as bureaucratic cogs in a machine. For instance, there is nothing which will endear you more with an administration than supplying it with *unsolicited* reports. This neat little trick has always worked wonders for me.

In such a case, prepare an itemized list of your expenses together with the corresponding receipts, join a copy of the program of the conference. The report proper, no more than two-page long, will describe precisely and specifically how the meeting was of benefit to you – what kind of ideas it gave you to test upon return to your laboratory, the gist of the information you acquired informally from fellow-conferees, …

A second, altogether different type is the story to be published in a newsletter or in a magazine. It will enjoy an audience of your fellow-scientists. In that case, it ought to consist of no more than ten **paragraphs** devoted each to one of the highlights of the conference. In selecting those key papers, try not to let yourself be influenced by what people around you were saying during the conference. Your report should be a reflection of your personal judgment not of groupthink. Try also not to let yourself become influenced by the eminence of the scientists who gave the papers. Aim only for what was quite objectively the best stuff.

The third type of conference report, which resembles somewhat the second, is also the most labor-intensive. In some conferences, one of the attendees, who was not one of the lecturers, is asked to prepare a summing-up, as the very last paper read at the conference. Such an exercise at synthesis, while an honor to your perceived impartiality and judgment, is a delicate undertaking.

These are the three most frequent traps: (i) out of courtesy, mentioning by name every single presenter; an instance of wrong-headed diplomacy, which will endear you to no one; (ii) not taking the time to put together a coherent account, and instead providing an incoherent mumble-jumble of undigested impressions; (iii) being long-winded. Remember: at the end of a conference, everyone is tired and just wants to go home.

Before you arrive at the location of the conference, prepare accordingly a questionnaire for your own use. Do not forget to bring it at the conference. Fill it as the conference proceeds, as a memo to yourself. Use it to prepare your text, no more than five-page long. This little bit of advance preparation works wonders.

REFERENCE: K. Blanchard and C. Root (2004) *Ready to write more: from paragraph to essay.* Prentice Hall, Lebanon IN.

positively brilliant 

CONSULTANT REPORT

You have just been hired by a company for your expert advice. Give answers to the questions which are put to you. You won't be able to do so from the top of your head. You will need to study the questions carefully. They won't be easy to answer, as a rule it will require hard work on your part. It also requires a great deal of documentation.

Where to find the information? Information provided by the client. Reports to stockholders from various companies. The Web. Newspapers and magazines. Books. Reports in libraries. Contacting people you know. This documentation stage is crucial, and it can be great fun. Double-check that the information you receive from various sources is consistent.

Once you have obtained the information, proceed to sort out the answers that were requested. In anticipation to meeting with company scientists, prepare the report you will be presenting, explaining and defending orally. It requires a table of contents, an **introduction** and a **conclusion**, a **bibliography**, diagrams and tables. Write it clearly and simply.

How does it differ from a publication in the open literature? There will be no need for an abstract; no need to make explicit the models or the tools for thought; no need for a discussion aimed at detailing your interpretation and at attempting to immunize it from criticism.

Make sure to answer each question put to you to the best of your ability. When you are unable to come up with an answer, say so and give the reasons why. Brace yourself for possible frustration, you may not be given any indication as to whether your answers and the proposals you suggested are deemed insufficient, just satisfactory, very helpful or positively brilliant. Your reward is two-fold: satisfaction from a job well done; plus financial compensation.

Do not kow-tow to what you perceive as the opinions and prejudices of your client. For one thing, the client won't present you with a single voice. You will have several interlocutors.

This means, conversely, that you may have to resist being drawn into the various power plays within the company. Remember: you are a hired outside expert, with an independent voice.

Do not reach beyond your area of expertise, nor overstep your mandate. You may have been hired simply to confirm in-house knowledge. Possibly, your client is using your expertise for defensive research to make sure that its existing proprietary process is not in risk of being overwhelmed by a competitor using the same kind of ideas you are presenting.

The challenge in this exercise is devising likely explanations for complex phenomena, on the spur of the moment. Finding a common language between pure and applied scientists can be a source of satisfaction. To evolve applications from newly-acquired knowledge even if unfamiliar to you, should also make you proud of having done something useful.

In return, you will receive a fee. The ballpark figure is between US k$ 1–5 per diem. This is negotiated beforehand. Be aware that the opening bid, within industrial circles, is often 50% of the maximum allowed figure.

A confidentiality agreement, by which you agree not to divulge any information is normally required. Stick to it strictly.

REFERENCE: R. Tepper (1987) *How to become a top consultant: how the experts do it.* Wiley, New York.

powerful incentives 

EXPERT TESTIMONY

Don't do it, unless … I start with the **conclusion**, so that my stand be clear from the outset. To volunteer your expertise into giving testimony in the courtroom runs against powerful negative incentives. Attorneys on the opposite side will do their utmost to destroy your credibility. They will hunt for dirt, for anything unsavory reflecting on your character and on your scientific integrity. During cross-examination, they will do their best to confront you with your own contradictions: they will get you to at least qualify your previous unambiguous assertions; they will confront them with the written record, at some cost to your credibility; any change of mind you may have had, in your published work, will be posted on the screen for everyone to see; they will gleefully post **sentences** you have penned which contradict one another. In short, they will try to make you look like a fool.

They will have invested their enormous resources in preparing to counter your arguments, which they will have had ample time to anticipate in their substance. They will have gone through your publications with a fine comb, recording in their computer archives a very large number of statements of yours bearing on the matter to be discussed at the trial. They will be well-prepared to destroy each assertion you have ever made in print. They will show to the judges contrary statements by authorities in the field, after making you admit publicly that indeed you regard Drs. X, Y, Z as leaders in the discipline. They will call their own experts to the stand, and those will testify under oath that your word is worthless. And the fiercest weapon in their arsenal of demolition will be your own contradictions: that he who has never self-contradicted cast the first stone …

For sure, there are powerful incentives for you to accept to give expert testimony. The three leading motivations, in order of decreasing importance, are the money, the adrenalin, and your personal ethics.

Compensation for testifying in court is mind-boggling, it can make consultancy fees look ridiculously puny; those are of the order of € 1,000 per hour at the time of writing (mid-2005). Tens of millions of dollars are at stake in big cases, involving for instance patent disputes between rival pharmaceutical companies. These are decided, to a large extent, on the basis of evidence commented upon by expert witnesses. If you are one of them, your part of the booty may be quite hefty.

In addition to being handsomely paid for being a hired gun, you may enjoy vicariously your day in court, the personal triumph of doing your best for your legal team to win, the rush of adrenalin from taking part in a heated debate and from imposing your views.

Last, your advocacy of the truth, of at least the truth of the matter in your eyes, is probably close to a moral imperative for you. It is part of your training as a scientist. Hence, you may have to dirty your hands and to agree to go to court, just as a matter of standing up for your beliefs, and of sharing the experience you have attained through hard work over many years.

First, do your homework. Many experts assume that because they are the world's leading authority on a topic, they simply have to show up and announce the right answer. They doubt that a mere lawyer could ever successfully challenge their testimony. That is a very serious mistake. The lawyer may not be an expert, but he or she has spent many months preparing for the moment of the cross-examination. The lawyer may not know all that the expert does – but the lawyer probably knows as much, or MORE than the expert about the narrow subject matter of the testimony. And the lawyer cares very, very much about winning the case. Therefore, he or she is prepared to take the expert through every assumption on which the testimony is based and ask extremely knowledgeable and thought-provoking questions. The expert who assumes that he knows more or can wing it, does so at his own peril.

Second, challenge the lawyers who have retained you. Most times, the lawyer-expert relationship is a good one. But sometimes there are some very bad facts in a case that a lawyer hides from the expert. Maybe he hopes that the expert will never find out, and that the expert will survive cross-examination because they won't come up. But if you don't want to look like a fool, press your »friendly« lawyer hard. What will the other side say? What evidence do they rely upon? Is there any evidence you haven't told me about? Most often, there aren't any secrets and these questions will just help you to be prepared. But sometimes there is the proverbial skeleton in the cupboard, and a probing question can save lots of embarrassment later.

Prepare also to recite your **biography**. The attorney who will run you through your testimony, as a witness on the stand will make you tell to the judge(s) who you are, before s/he puts questions to you. The team of attorneys you will be working in will coach you for your whole testimony, starting with the details about your professional career. What would not be helpful, generally speaking, is for you to assume on the stand too arrogant and too cocky an attitude. Present yourself at least as much as a servant of the truth than as a spokesman for this deity, with a direct line to her.

Your friends the attorneys will run you through your testimony, in rehearsal sessions. The attorneys on the opposite side, during cross-examination, will put questions to you too. You will have been coached for this purpose. But all the aggressive questions won't have been anticipated. Give answers which are brief and clear. Be unambiguous, if you can.

Grenades and artillery shells will be thrown at you, metaphorically-speaking. They are meant to shatter your composition, to have you lose credibility because you will have been made emotional. Do not raise your voice, remain calm, continue talking in the same soft, gentle manner, at the same leisurely pace. Almost certainly, you will be asked questions you don't know the answers to. You will have to say so, you are not a Dr. Know-it-all. However, you may undo your whole testimony by confessing your ignorance too often. To state it once or twice will only raise your stature, because you will come across as a responsible individual.

REFERENCE: A. A. Moonssens, J. E. Starrs and C. E. Henderson (1995) *Scientific evidence in civil and criminal cases,* 4th edn, Foundation Press, New York.

GENERAL BIBLIOGRAPHY, SELECTED

Arnheim R. (1969) Visual Thinking. University of California Press, Berkeley CA

Avramov I. (1999) An apprenticeship in scientific communication: the early correspondence of Henry Oldenburg (1656–63). Notes and Records of the Royal Society of London 53(2):187–201

Battalio, J.T. (1998) The rhetoric of science in the evolution of American ornithological discourse. Ablix, Stamford, CT

Berkenkotter C., Huckin T. (1995) Genre knowledge in disciplinary communication: cognition/culture/power. Erlbaum, Hillsdale, NJ

Briscoe M.H. (1996) Preparing Scientific Illustrations. 2nd ed. Springer, New York NY.

Bucchi M. (1996) When scientists turn to the public: alternative routes in science communication. Public Understanding of Science 5:375–394

Bucchi M. (1998) Science and the media: alternative routes in scientific communication. Routledge, London

Darian S. (2003) Understanding the language of science. University of Texas Press, Austin, TX

Davies D. (1990) The telling image: the changing balance between pictures and words in a technological age. Oxford University Press, Oxford

Day S.B., (ed) (1975) Communication of scientific information. Karger, Basel

Dunwoody S. (1980) The science writing inner club: a communication link between science and the lay public. Science, Technology and Human Values 30:14–22

Ford B.J. (1993) Images of science: a history of scientific illustration. Oxford University Press, Oxford

Frankel F. (2002) Envisioning Science. The design and craft of the science image. MA: The MIT Press, Cambridge

Friedman S.M., Dunwooly S., Rogers C. L. (eds) (1999) Communicating uncertainty: media coverage of new and controversial science. Erlbaum, Mahwah, NJ

Goodfield J. (1981) Reflections on science and the media. American Association for the Advancement of Science, Washington DC

Gregory, J. (1998) Science in public: Communication, culture and credibility. Plenum Trade, New York

Gross A.G., Harmon J.E., Reidy M. (2002) Communicating science: the scientific article from the seventeenth century to the present. Oxford University Press, Oxford

Hager P.J., Scheiber H.J. (1997) Designing &Delivering Scientific, Technical, and Managerial Presentations. Wiley-Intersience, New York NY.

Hall M.B. (1975) The Royal Society's role in the diffusion of information in the 17th century. Notes and Records of the Royal Society of London 29:173–192

Hall B.S. (1996) The Didactic and the Elegant: some thoughts on scientific and technological illustrations in the middle ages and the renaissance. IN: Picturing knowledge: historical and philosophical problems concerning the use of art in science, 3–39. University of Toronto Press, Toronto

Johns A. (1998) The Nature of the book: print and knowledge in the making. University of Chicago Press, Chicago

Jones G., Connell I., Meadows J. (1978) The presentation of science by the media. University of Leicester, Primary Communication Research Centre, Leicester

Kaunzner W. (1987) On the transmission of mathematical knowledge in Europe. Sudhoffs Archiv 71:129–140

Lenoir T. (ed) (1998) Inscribing science: scientific texts and the materiality of communication. Stanford University Press, Palo Alto CA

Lewenstein B.V. (1995) From fax to facts: Communication in the cold fusion saga. Social Studies of Science 25:403–436

Lilley D.B. (1989) A history of information science, 1945–1985. Academic Press, San Diego, CA

Macdonald S. (1992) Science on display: the representation of scientific controversy in museum exhibitions. Public Understanding of Science 1:69–87

Malaquias I.M. and Thomaz M.F. (1994) Scientific communication in the 18th century. The case of John Hyacinth de Magellan. Physis: Rivista Internazionale di Storia della Scienze 31:817–834

Manten A.A. (1980) Publication of scientific information is not identical with communication. Scientometrics 2:303–308

Martin J.R. and Veel R. (eds) (1998) Reading science: critical and functional perspectives on discourses of science. Routledge, London

Meadows A.J. (1974) Communication in Science. Butterworths, London

Montgomery S.L. (2003) The Chicago guide to communicating science. University of Chicago Press, Chicago

Munk-Jorgensen P. (2003) The privilege of editing a scientific journal. Epidemiologia e Psichiatria Sociale 12(1):2–4

Aborn M. (ed) (1988) Telescience: scientific communication in the information age. Ann Amer Acad Pol Soc Sci 495:10–143

Nadel, E. (1980) Formal communication, journal concentration, and the rise of a discipline in physics. Sociology 14:401–416

Neeley K. A. (1992) Women as Mediatrix: women as writers on science and technology in the eighteenth and nineteenth centuries. IEEE Transactions on Professional Communication 35:208–216

Nelkin D. (1995) Selling science: how the press covers science and technology. W. H. Freeman, New York

Parker E. B., Paisley W. J., and R. Garret. (1967) Bibliographic citations as unobtrusive measures of scientific communication. Stanford University, Institute for Communication Research, Palo Alto, CA

Pastore G. (1990) Communication prerequisites in the science of imaging. Rays 15(1): 25–32; 99–100

Pera M. (1994) The discourses of science. University of Chicago Press, Chicago

de Solla Price D. (1980) On the scientific element in a scientific communication. Interciencia 5:220–222

Rossiter M. (1986) Women and the history of scientific communication. Journal of Library History 21:39–59

Simpson C. (1994) Science of coercion: communication research and psychological warfare, 1945–1960. Oxford University Press, New York

de Sousa R. C. (1990) La vulgarisation scientifique: devoir et plaisir dans la communication du savoir. Cahiers de la Faculté de Médecine (N° 19):81–112

Tufte E. R. (1990) Envisioning Information. Graphics Press, Cheshire, CT

Turabian K. L. (1996) A manual for writers of term papers, theses and dissertations. 6th. edn., Translated by revised by John Grossman and Alice Bennett. Chicago Guides to Writing, Editing and Publishing, University of Chicago Press, Chicago

Tyrer P. (2003) Entertaining eminence in the *British Journal of Psychiatry*. The British Journal of Psychiatry 183:1–2

Wilkinson G. (2003) Editing the *British Journal of Psychiatry*. Epidemiologia e Psichiatria Sociale 12(1):5–8

Williams J. J. (1998) Editorial instinct: an interview with William P. Germano. The Minnesota Review

Zott R. (1993) The development of science and scientific communication: Justus v. Liebig's two famous publications of 1840. Ambix 40:1–10

INDEX, CONCEPTUAL

INDEX, PRACTICAL